Introduction to Floriculture

Introduction to Floriculture

Caroline Harrington

Larsen & Keller
www.larsen-keller.com

Introduction to Floriculture
Caroline Harrington
ISBN: 978-1-64172-145-5 (Hardback)

Larsen & Keller
Published by Larsen and Keller Education,
5 Penn Plaza,
19th Floor,
New York, NY 10001, USA

Cataloging-in-Publication Data

Introduction to floriculture / Caroline Harrington.
p. cm.
Includes bibliographical references and index.
ISBN 978-1-64172-145-5
1. Floriculture. 2. Ornamental horticulture. I. Harrington, Caroline.
SB404.9 .I58 2019
635.9--dc23

Printed and bound in China.

For more information regarding Larsen and Keller Education and its products, please visit the publisher's website www.larsen-keller.com

Table of Contents

Preface

Floriculture is a branch of horticulture that is concerned with the cultivation of flowering and ornamental plants. Crops of interest in floriculture include bedding plants, houseplants, flowering garden plants, pot plants, etc. Flowering plants are mostly sold in pots for decoration in homes. Some of the common breeds of flowering plants are poinsettias, orchids, chrysanthemums, azaleas, etc. Cut flowers are sold as bouquets. Farming of flowers and foliage employs special techniques of floriculture, such as spacing, training and pruning for optimal flower harvest, as well as chemical treatments for storage, preservation and packaging after harvest. This book unfolds the innovative aspects of floriculture, which will be crucial for the holistic understanding of the subject matter. The topics included in this textbook are of the utmost significance and bound to provide incredible insights to readers. Coherent flow of topics, student-friendly language and extensive use of examples make this book an invaluable source of knowledge.

To facilitate a deeper understanding of the contents of this book a short introduction of every chapter is written below:

Chapter 1- Floriculture falls under the domain of horticulture. It is concerned with the study of flowering and ornamental plants along with their cultivation. Such plants are grown in pots, cell packs or hanging baskets. They are considered vital for gardening and landscaping. This is an introductory chapter, which will introduce briefly all the significant aspects of floriculture.

Chapter 2- A flower is the most crucial element of plant reproduction. This chapter discusses in detail the structure of a flower. It elucidates the core topics of plant reproductive morphology, inflorescence and flower development. It has been carefully written to provide an understanding of the varied aspects of the anatomy and biology of flowers.

Chapter 3- Pollination is the transfer of pollen from the male part of a plant to a female part. It enables fertilization and the consequent production of seeds. The aim of this chapter is to explore the different mechanisms of pollination and fertilization in flowers.

Chapter 4- Flowering plants form the most diverse group among land plants with over two hundred thousand known species. Such plants are widely grown for aesthetic appeal. Poinsettias, chrysanthemums, orchids and azaleas are commonly cultivated flowering plants. The topics elaborated in this chapter will help in gaining a better perspective about the plants that are of significance in floriculture.

Chapter 5- Many varieties of flowering trees are planted for aesthetic appeal. Some important examples are cherry blossom, magnolia, delonix regia, etc. The Japanese art form of bonsai is another significant practice that is intended for decoration of indoor and outdoor settings. This chapter closely examines the techniques and principles of practicing bonsai. It also discusses some common flowering trees that can be cultivated in floriculture.

Chapter 6- Flowering plants are susceptible to a wide range of diseases, which have been discussed in detail in this chapter. Cymbidium mosaic virus, Diplocarpon rosae and Diaporthe helianthi are some of the pathogens that cause diseases in flowering plants such as orchids, roses and sunflowers. This chapter also discusses in detail the management of these diseases in flowers.

I owe the completion of this book to the never-ending support of my family, who supported me throughout the project.

Caroline Harrington

1

Floriculture: An Introduction

Floriculture falls under the domain of horticulture. It is concerned with the study of flowering and ornamental plants along with their cultivation. Such plants are grown in pots, cell packs or hanging baskets. They are considered vital for gardening and landscaping. This is an introductory chapter, which will introduce briefly all the significant aspects of floriculture.

Floriculture

Floriculture, or flower farming, is a discipline of horticulture concerned with the cultivation of flowering and ornamental plants for gardens and for floristry, comprising the floral industry. The development, via plant breeding, of new varieties is a major occupation of floriculturists.

Floriculture crops include bedding plants, houseplants, flowering garden and pot plants, cut cultivated greens, and cut flowers. As distinguished from nursery crops, floriculture crops are generally herbaceous. Bedding and garden plants consist of young flowering plants (annuals and perennials) and vegetable plants. They are grown in cell packs (in flats or trays), in pots, or in hanging baskets, usually inside a controlled environment, and sold largely for gardens and landscaping. *Pelargonium* ("geraniums"), *Impatiens* ("busy lizzies"), and *Petunia* are the best-selling bedding plants. The many cultivars of *Chrysanthemum* are the major perennial garden plant in the United States.

Flowering plants are largely sold in pots for indoor use. The major flowering plants are poinsettias, orchids, florist chrysanthemums, and finished florist azaleas. Foliage plants are also sold in pots and hanging baskets for indoor and patio use, including larger specimens for office, hotel, and restaurant interiors.

Cut flowers are usually sold in bunches or as bouquets with cut foliage. The production of cut flowers is specifically known as the cut flower industry. Farming flowers and foliage employs special aspects of floriculture, such as spacing, training and pruning plants for optimal flower harvest; and post-harvest treatment such as chemical treatments, storage, preservation and packaging. In Australia and the United States some species are harvested from the wild for the cut flower market.

Bedding (Horticulture)

Bedding, in horticulture, refers to the temporary planting of fast-growing plants into flower beds to create colourful, temporary, seasonal displays, during spring, summer or winter. Plants used for bedding are generally annuals, biennials or tender perennials; succulents are gaining in popularity.

Some bedding plants are also referred to as "patio plants" because they are widely used in pots and other containers positioned on patios, terraces, decking and other areas around houses. Larger tender "conservatory plants" may also be moved out from greenhouses or conservatories and planted out in borders (or stood in their pots in sheltered positions) for the warmer months, then returned to shelter for the winter.

The modern bedding plant industry breeds and produces plants with a neat, dwarf habit, which flower uniformly and reliably. They are bred primarily for use in large-scale bedding schemes where uniformity and predictability is of paramount importance, but this is often achieved by losing the plants' individual character, and has been criticised by such notable plantsmen as the late Christopher Lloyd, who championed an informal style of bedding.

Spring bedding with tulips, wallflowers and forget-me-nots

Bedding Plants

There exists a huge range of plants specifically grown to produce a period of flower colour throughout the spring and summer, and (usually) discarded after flowering. They may conveniently be divided into four groups:-

- Hardy annuals sown directly into the ground early in the season (poppy, stock, sunflower, clarkia, godetia, eschscholzia, nigella, dianthus)
- Tender annual or perennial plants treated as half-hardy annuals - sown under glass in late winter in heat, or purchased as young plants, and hardened-off outdoors when all danger of frost has passed (begonia, lobelia, petunia, argyranthemum, chrysanthemum, pelargonium, nicotiana, cosmos, fuchsia)
- Hardy biennial plants, or perennials treated as biennial, sown in one year to

flower the next, and discarded after flowering (antirrhinum, polyanthus, wallflower, daisy, foxglove, some dianthus, some poppies, campanula, delphinium, aubrieta, aquilegia, cornflower, pansies)

- Corms, rhizomes, bulbs and tubers, planted each year and lifted after the plant has died down and stored in winter, or discarded (tulip, narcissus, hyacinth, gladiolus, dahlia, canna)

Types of Bedding

Formal bedding, as seen in parks and large gardens, where whole flower beds are replanted two or three times a year, is a costly and labour-intensive process. Towns and cities are encouraged to produce impressive displays by campaigns such as "Britain in bloom".

Spring bedding (tulips and polyanthus), South Shields, UK

Spring Bedding

Plants used for spring bedding are often biennials (sown one year to flower the next), or hardy, but short-lived, perennials. Spring-flowering bulbs such as tulips are often used, typically with forget-me-nots, wallflowers, winter pansies and polyanthus.

Summer Bedding

Formal summer bedding scheme in Victoria Park, Bath, Somerset. Most of the colour is provided by large-flowered begonias

Plants used for summer bedding are generally annuals or tender perennials. They become available (often as what are referred to as "plug plants") in nurseries and garden centres during spring, to be gradually "hardened off" (acclimatised to outdoor conditions) by the purchaser and finally planted out around the time that the last frosts are expected. Experienced gardeners keep an eye on the weather forecasts at that time of year and are on standby to protect their bedding displays overnight with horticultural fleece (or the older alternatives of net curtains or newspaper) if frost threatens.

Carpet Bedding

Municipal carpet bedding schemes often incorporate lettering

Carpet bedding employs two or more contrasting plant cultivars with a neat, dwarf habit and distinct colouring (of flower or foliage) to create geometric displays. It is often used to form such things as lettering, logos or trademarks, coats of arms, or floral clocks. Suitable plants are rosette-forming succulents such as *Echeveria* or fairly slow-growing or mat-forming foliage plants, such as coloured-leaved *Alternanthera* cultivars, which are tolerant of clipping; such plants may also be used in three-dimensional sculptural forms or pseudo-topiary.

Winter Bedding

Planted in autumn to give a display until early spring, the plants used for winter bedding are mainly hardy perennials. As it has to be planted at the same time of year as spring bedding does, winter bedding tends to be less commonly seen, except in containers such as windowboxes. Some are short-lived and will be discarded after their first display; others may be used as a source of cuttings for the next year. Winter-hardy ornamental vegetables such as cultivars of kale and cabbage with coloured or variegated foliage are increasingly common. *Primula* cultivars (polyanthus and primroses) are commonly used, as are winter-flowering heathers and *Viola ×wittrockiana*, winter pansies. Variegated evergreens such as cultivars of *Vinca minor* (lesser periwinkle), *Euonymus fortunei* and *Hedera helix* (ivies) are also popular.

Flower Pot Farming

Pot farming, flowerpot farming or pot agriculture is the practice of growing flower plants, exclusively in containers (flowerpots), instead of planting them in the ground.

The balcony, backyard, patio or even the doorstep can all be turned over to vegetable production.

Species

It is important to select appropriate species to ensure success in the urban garden. Species such as parsley, tomato, spinach or arugula can even be grown inside the house, if there is adequate light and ventilation. In outdoor terraces, options include basil, rosemary, thyme, dill, oregano, carrots, radish, anise, cidron, mint, among others. Special care must be taken to grow these plants in pots. This practice is associated with trends as Zero Kilometers, where the consumption of local vegetables is encouraged, to reduce carbon emissions related transporting food.

Flowerpot

A terracotta flowerpot

Jiffy pots: peat pots that are biodegradable and may be planted directly into the soil

Victorian decorative flowerpots at Kindrogan House, Enochdu, Scotland

A flowerpot, flower pot, or plant pot is a container in which flowers and other plants are cultivated and displayed. Historically, and still to a significant extent today, they are made from terracotta. Flowerpots are now often also made from plastic, wood, stone, or sometimes biodegradable material. An example of biodegradable pots are ones made of heavy brown paper, cardboard, or peat moss in which young plants for transplanting are grown.

For seedling starting in commercial greenhouses or polytunnels, pots usually take the form of trays with cells, each cell acting as one small pot. These trays are often called flats.

There are often holes in the bottom of pots, to allow excess water to flow out, sometimes to a saucer that is placed under the flowerpot. The plant can use this water with its roots, as needed. Recently, some flowerpots have been made with an automatic watering system, using a reservoir.

Flowerpots

Flowerpots have a number of uses such as transporting plants to new locations, starting seeds, patio and indoor cultivation of plants, and the growing of tender plants in colder regions indoors. Through the centuries, the use of flowerpots has influenced the horticultural use of plants, and the Egyptians were among the first to use pots to move plants from one location to another. The Romans brought potted plants inside during cold weather. Pots were used to ship breadfruit seedlings from Tahiti to the West Indies. Also Orchids, African violets and *Pelargonium* geraniums were shipped in pots from other parts of the world, including Africa, to North America and Europe.

Josiah Wedgwood's flowerpots were as popular as his famous dinner-ware, they were often highly decorative and used as table centrepieces.

In Athens, earthenware flowerpots were thrown into the sea during the festival of the Gardens of Adonis. Theophrastus, c. 371 – c. 287 BC, mentions that a plant called southern-wood was raised and propagated in pots because it was difficult to grow.

The top of the flowerpot underneath the rim is commonly known as the shoulder or collar and can aid handling.

Traditional Classification

Flower pots were traditionally made from terracotta. They were made and sold by the cast, which is the number of pots produced from a given quantity of clay.

The traditional sizes were as follows, although others existed:

Name	No. to cast	Top diameter (inches)	Depth (inches)
Ones	1	20	18
Twos	2	18	14
Fours	4	15	13
Sixes	6	13	12
Eights	8	12	11
Twelves	12	11.5	10
Sixteens	16	9.5	9
Twenty-fours	24	8.5	8
Thirty-twos	32	6	6
Forty-eights	48	4.5	5
Sixties	60	3	3.5
Seventy-twos or thimbles	72	2.5	
Thumbs	80	2.5	2.5
Nineties or thumbs	90	1	

Other sources give different values, and sometimes names, for the smaller pots, for example *The Gardener's Everyday Log Book* while agreeing on "twos" to "thirty-twos" has two different types of "sixties" and disagrees on "thumbs" and "thimbles":

Name	Top diameter (inches)	Depth (inches)
Large sixties	4	3.5
Small sixties	3	3
Thumbs	2.5	2.5
Thimbles	2	2

A taller and thinner shape of pot, suitable for deep-rooting plants, was known as a long tom, a term still used. The traditional size for a long tom used for auriculas was 3 in diameter by 3.75 to 4 in depth.

Nursery Pots

In the nursery business, plants are generally grown in round or square plastic pots.

United States

The sizes of plastic pots have been assigned an ANSI standard by the American Nursery and Landscape Association. Pots designated #1–#100 nominally have the volume of that many gallons, but in fact a #1 pot has a capacity of 0.625 gallons (a "trade gallon"). There is also a Small Plant series: SP1, 6.5–8.0 in^3; SP2, 13.0–15.0 in^3; SP3, 20.0–30.0 in^3; SP4, 51–63 in^3; SP5, 93–136 in^3. An SP4 pot is commonly called a "4-inch" or "quart" container.

Europe

Plastic pots come in a number of standard sizes, with a code indicating the approximate dimensions of the diameter at the top.

Code	Diameter	Height	Volume	Approx. Imperial equivalent
6F	6 cm	6 cm		2.25"
8F	8 cm	7.5 cm		3"
9F	9 cm	9 cm		3.5"
10F	10 cm	9.5 cm	0.5 l	4"
13F/14A	13 cm	12 cm	0.9 l	5.5"
15F	15 cm	14 cm		6"

Window Box

A window box (sometimes called a window flower box, or window box planter) is a type of flower container for live flowers or plants in the form of a box attached on or just below the sill of a window on a building or house. It may also be used for growing herbs or other edible plants.

Window boxes in Valparaiso, Chile

Description

A window box is usually placed on a window sill or fixed to the wall immediately below it so the owner(s) can easily get to the plants in the container. If a window box is installed under a window it is usually supported in place by brackets on the wall below. Some materials, such as PVC or fibre glass, use a cleat mounting system from behind to attach to the building or may be directly bolted to the home without the use of support brackets below.

Wood, brick, terracotta, metal, fibre glass, vinyl, and cellular PVC may all be used in window box construction. A typical wooden window box will last 3–5 years before showing deterioration, though with painting and maintenance they can last 10–15 years. Fibre glass and plastic may be used because they are cheap and lightweight.

Window boxes are often used by people who live on the upper floors of flats or apartments, and thus do not have access to a garden or patio to grow flowers. Window boxes enable the plants to be readily seen by those inside the property as well as outside. Larger boxes, 10–12 inches in height, can be used to plant items that need more root space and to allow flowers and plants to be layered in multiple rows to create more intricate flower displays. Access for planting and maintenance can be via the window from indoors.

Flower Box

A flower box is a type of container in the form of a planter or box that is usually placed outdoors and used for displaying live plants and flowers, but it may also be used for growing herbs or other edible plants.

It is usually placed or affixed to an accessible location so the resident of a home may easily work with the plants in the container. A flower box may be installed under a window and supported in place by brackets on the wall below, in which case it may be called a window box. Flower boxes may also be used to line decks, patios, porches, steps, and sidewalks and they can even be hung from railings.

Flower box display at the Chelsea Flower

Wood, brick, metal, fiberglass, vinyl, and cellular PVC can all be used in flower box construction, with wood being a classical material of choice. A typical wooden container will last 3–5 years before showing signs of rot. With painting and maintenance they can sometimes last 10–15 years. Fiberglass has the advantage of being lightweight and insect proof. Vinyl and cellular PVC are plastics which are completely rot proof alternatives to wood which are often used on homes to prevent rot or siding damage.

Sometimes a box is placed inside a kitchen window in order to grow herbs or other supplies for a chef as an easily accessed miniature kitchen garden.

J. Linderski has argued that Pliny described flower boxes in his *Naturalis Historia*, at 19.59. However, Linderski could only find one other allusion to this practice in Martial 11.18.

Flower Garden

Flower garden at the Botanic Gardens, Churchtown, Southport, Merseyside, England

Flower garden in Norway, Arboretum in Bergen

Flower gardens combine plants of different heights, colors, textures, and fragances to create interest and delight the senses.

A flower/floral garden is any garden where flowers are grown and displayed.

Because flowers bloom at varying times of the year, and some plants are annual, dying each winter, the design of flower gardens can take into consideration maintaining a sequence of bloom and consistent color combinations through varying seasons. Besides organizing the flowers in bedding-out schemes limited to annual and perennial flower beds, careful design also takes the labour time, and the color pattern of the flowers into account.

The labour time can be decreased by using techniques such as mulching, companion planting, use of specific flowers/plants suppressing grass, ... in the vicinity (i.e. Rhinanthus, ...).

Flower color is another important feature of both the herbaceous border and the mixed border that includes shrubs as well as herbaceous plants. Flower gardens are sometimes tied in function to other kinds of gardens, like knot gardens or herb gardens, many herbs also having decorative function, and some decorative flowers being edible.

A simpler alternative to the designed flower garden is the "wildflower" seed mix, with assortments of seeds which will create a bed that contains flowers of various blooming seasons, so that some portion of them should always be in bloom. The best mixtures even include combinations of perennial and biennials, which may not bloom until the following year, and also annuals that are "self-seeding", so they will return, creating a permanent flowerbed.

Another, even more recent trend is the "flower garden in a box", where the entire design of a flower garden is pre-packaged, with separate packets of each kind of flower, and a careful layout to be followed to create the proposed pattern of color in the garden-to-be.

Cutting garden

A functional garden used to grow flowers for indoor use rather than outdoor display is known as a cutting garden. It is usually only a feature of large residences.

The cutting garden is typically placed in a fertile and sunlight position out of public view and is not artistically arranged, as it contains flowers for *cutting*. The cutting garden may comprise a herb garden and ornamental vegetables as well.

Flowering Plant

The flowering plants, also known as angiosperms, Angiospermae or Magnoliophyta, are the most diverse group of land plants, with 416 families, approximately 13,164 known genera and c. 295,383 known species. Like gymnosperms, angiosperms are seed-producing plants. However, they are distinguished from gymnosperms by characteristics including flowers, endosperm within the seeds, and the production of fruits that contain the seeds. Etymologically, *angiosperm* means a plant that produces seeds within an enclosure; in other words, a fruiting plant. The term comes from the Greek words *angeion* ("case" or "casing") and *sperma* ("seed").

The ancestors of flowering plants diverged from gymnosperms in the Triassic Period, 245 to 202 million years ago (mya), and the first flowering plants are known from 160 mya. They diversified extensively during the Lower Cretaceous, became widespread by 120 mya, and replaced conifers as the dominant trees from 100 to 60 mya.

Reproduction

Fertilization and Embryogenesis

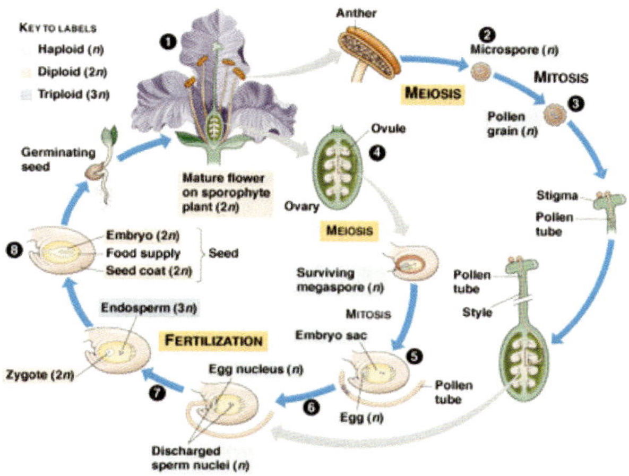

Angiosperm life cycle

Double fertilization refers to a process in which two sperm cells fertilize cells in the ovule. This process begins when a pollen grain adheres to the stigma of the pistil (female reproductive structure), germinates, and grows a long pollen tube. While this pollen tube is growing, a haploid generative cell travels down the tube behind the tube nucleus. The generative cell divides by mitosis to produce two haploid (n) sperm cells. As the pollen tube grows, it makes its way from the stigma, down the style and into the ovary. Here the pollen tube reaches the micropyle of the ovule and digests its way into one of the synergids, releasing its contents (which include the sperm cells). The synergid that the cells were released into degenerates and one sperm makes its way to fertilize the egg cell, producing a diploid ($2n$) zygote. The second sperm cell fuses with both central cell nuclei, producing a triploid ($3n$) cell. As the zygote develops into an embryo, the triploid cell develops into the endosperm, which serves as the embryo's food supply. The ovary will now develop into a fruit and the ovule will develop into a seed.

Fruit and Seed

As the development of embryo and endosperm proceeds within the embryo sac, the sac wall enlarges and combines with the nucellus (which is likewise enlarging) and the integument to form the *seed coat*. The ovary wall develops to form the fruit or pericarp, whose form is closely associated with type of seed dispersal system.

The fruit of the *Aesculus* or Horse Chestnut tree

Frequently, the influence of fertilization is felt beyond the ovary, and other parts of the flower take part in the formation of the fruit, e.g., the floral receptacle in the apple, strawberry, and others.

The character of the seed coat bears a definite relation to that of the fruit. They protect the embryo and aid in dissemination; they may also directly promote germination. Among plants with indehiscent fruits, in general, the fruit provides protection for the embryo and secures dissemination. In this case, the seed coat is only slightly developed. If the fruit is dehiscent and the seed is exposed, in general, the seed-coat is well developed, and must discharge the functions otherwise executed by the fruit.

Meiosis

Flowering plants generate gametes using a specialized cell division called meiosis. Meiosis takes place in the ovule (a structure within the ovary that is located within the pistil at the center of the flower). A diploid cell (megaspore mother cell) in the ovule undergoes meiosis (involving two successive cell divisions) to produce four cells (megaspores) with haploid nuclei. One of these four cells (megaspore) then undergoes three successive mitotic divisions to produce an immature embryo sac (megagametophyte) with eight haploid nuclei. Next, these nuclei are segregated into separate cells by cytokinesis to producing 3 antipodal cells, 2 synergid cells and an egg cell. Two polar nuclei are left in the central cell of the embryo sac.

Pollen is also produced by meiosis in the male anther (microsporangium). During meiosis, a diploid microspore mother cell undergoes two successive meiotic divisions to produce 4 haploid cells (microspores or male gametes). Each of these microspores, after further mitoses, becomes a pollen grain (microgametophyte) containing two haploid generative (sperm) cells and a tube nucleus. When a pollen grain makes contact with the female stigma, the pollen grain forms a pollen tube that grows down the style into the ovary. In the act of fertilization, a male sperm nucleus fuses with the female egg nucleus to form a diploid zygote that can then develop into an embryo within the newly forming seed. Upon germination of the seed, a new plant can grow and mature.

The adaptive function of meiosis is currently a matter of debate. A key event during meiosis in a diploid cell is the pairing of homologous chromosomes and homologous recombination (the exchange of genetic information) between homologous chromosomes. This process promotes the production of increased genetic diversity among progeny and the recombinational repair of damages in the DNA to be passed on to progeny. To explain the adaptive function of meiosis in flowering plants, some authors emphasize diversity and others emphasize DNA repair.

Apomixis

Apomixis (reproduction via asexually formed seeds) is found naturally in about 2.2% of angiosperm genera One type of apomixis, gametophytic apomixis found in a dandelion species involves formation of an unreduced embryo sac due to incomplete meiosis (apomeiosis) and development of an embryo from the unreduced egg inside the embryo sac, without fertilization (parthenogenesis).

Uses

Agriculture is almost entirely dependent on angiosperms, which provide virtually all plant-based food, and also provide a significant amount of livestock feed. Of all the families of plants, the Poaceae, or grass family (providing grains), is by far the most important, providing the bulk of all feedstocks (rice, maize, wheat, barley, rye, oats, pearl millet, sugar cane, sorghum). The Fabaceae, or legume family, comes in second place. Also of high importance are the Solanaceae, or nightshade family (potatoes, tomatoes, and peppers, among others); the Cucurbitaceae, or gourd family (including pumpkins and melons); the Brassicaceae, or mustard plant family (including rapeseed and the innumerable varieties of the cabbage species *Brassica oleracea*); and the Apiaceae, or parsley family. Many of our fruits come from the Rutaceae, or rue family (including oranges, lemons, grapefruits, etc.), and the Rosaceae, or rose family (including apples, pears, cherries, apricots, plums, etc.).

In some parts of the world, certain single species assume paramount importance because of their variety of uses, for example the coconut (*Cocos nucifera*) on Pacific atolls, and the olive (*Olea europaea*) in the Mediterranean region.

Flowering plants also provide economic resources in the form of wood, paper, fiber (cotton, flax, and hemp, among others), medicines (digitalis, camphor), decorative and landscaping plants, and many other uses. The main area in which they are surpassed by other plants — namely, coniferous trees (Pinales), which are non-flowering (gymnosperms) — is timber and paper production.

Ornamental Plant

Ornamental plants are plants that are grown for decorative purposes in gardens and landscape design projects, as houseplants, for cut flowers and specimen display. The cultivation of these, called floriculture, forms a major branch of horticulture.

Ornamental petunia plant

Garden Plants

Commonly, ornamental [garden] plants are grown for the display of aesthetic features including: flowers, leaves, scent, overall foliage texture, fruit, stem and bark, and aesthetic form. In some cases, unusual features may be considered to be of interest, such as the prominent thorns of *Rosa sericea* and *cacti*. In all cases, their purpose is for the enjoyment of gardeners, visitors, and the public institutions.

Trees

Similarly certain trees may be called ornamental trees. This term is used when they are used as part of a garden, park, or landscape setting, for instance for their flowers, their texture, form ,size and shape, and other aesthetic characteristics. In some countries trees in 'utilitarian' landscape use such as screening, and roadside plantings are called *amenity trees*.

Grasses

Ornamental grasses are grasses grown as ornamental plants. Many ornamental grasses are true grasses (Poaceae), however several other families of grass-like plants are typically marketed as ornamental grasses. These include the sedges (Cyperaceae), rushes (Juncaceae), restios (Restionaceae), and cat-tails (Typhaceae). All are monocotyledons, typically with narrow leaves and parallel veins. Most are herbaceous perennials, though many are evergreen and some develop woody tissues. Ornamental grasses are popular in many They bring striking linear form, texture, color, motion, and sound to the garden, throughout the year.

Ornamental grasses are popular in many colder hardiness zones for their resilience to cold temperatures and aesthetic value throughout fall and winter seasons.

Cultivation

Meillandine Rose in clay pot

For plants to be considered to be ornamental, they may require specific work and activity by a gardener. For instance, many plants cultivated for topiary and bonsai would only be considered to be ornamental by virtue of the regular pruning carried out on them by the gardener, and they may rapidly cease to be ornamental if the work was abandoned.

Ornamental plants and trees are distinguished from utilitarian and crop plants, such as those used for agriculture and vegetable crops, and for forestry or as fruit trees. This does not preclude any particular type of plant being grown both for ornamental qualities in the garden, and for utilitarian purposes in other settings. Thus lavender is typically grown as an ornamental plant in gardens, but may also be grown as a crop plant for the production of lavender oil.

The Term

The term ornamental plant is used here in the same sense that it is generally used in the horticultural trades. The term largely corresponds to 'garden plant', though the latter is much less precise, as any plant may be grown in a garden. Ornamental plants are plants which are grown for display purposes, rather than functional ones. While some plants are both ornamental and functional, people usually use the term "ornamental plants" to refer to plants which have no value beyond being attractive, although many people feel that this is value enough. Ornamental plants are the keystone of ornamental gardening, and they come in a range of shapes, sizes and colors suitable to a broad array of climates, landscapes, and gardening needs.

Some ornamental plants are grown for showy foliage. Their foliage may be deciduous, turning bright orange, red, and yellow before dropping off in the fall, or evergreen, in which case it stays green year-round. Some ornamental foliage has a striking appearance created by lacy leaves or long needles, while other ornamentals are grown for distinctively colored leaves, such as silvery-gray ground covers and bright red grasses, among many others.

Other ornamental plants are cultivated for their blooms. Flowering ornamentals are a key aspect of many gardens, with many flower gardeners preferring to plant a variety of flowers so that the garden is continuously in flower through the spring and summer. Depending on the types of plants being grown, the flowers may be subtle and delicate, or large and showy, with some ornamental plants producing distinctive aromas which paint a palette of scents in addition to colors.

Ornamental Grass

Crimson fountaingrass (*Pennisetum setaceum*)

Ornamental grasses are grasses grown as ornamental plants. They have become increasingly popular in gardens in recent years. Ornamental grasses are popular in many colder hardiness zones for their resilience to cold temperatures and aesthetic value throughout fall and winter seasons.

Classifications

Along with true grasses (Poaceae), several other families of grass-like plants are typically marketed as ornamental grasses. These include the sedges (Cyperaceae), rushes (Juncaceae), restios (Restionaceae), and cat-tails (Typhaceae). All are monocotyledons, typically with narrow leaves and parallel veins. Most are herbaceous perennials, though many are evergreen and some develop woody tissues. They bring striking linear form, texture, color, motion, and sound to the garden, throughout the year.

Habits

Almost all ornamental grasses are perennials, coming up in spring, from their roots, which have stored large quantities of energy, and in fall or winter go dormant. Some, notably bamboos, are evergreen, and a few are annuals. Many are bunch grasses and tussock grasses, though others form extensive systems of many-branched rhizomes. The bunching types are often called "clump-forming" or "clumping", distinct from the rhizomatous types, called "running". Sizes vary from a few centimeters up to several meters; the larger bamboos may reach 20 m or more tall. Some ornamental grasses are species that can be grown from seed. Many others are cultivars, and must be propagated by vegetative propagation of an existing plant.

Pampas grass (*Cortaderia selloana*) is easily recognizable, with semi-dwarf to very large selections for the landscape. Deer grass (*Muhlenbergia rigens*) and canyon prince wild blue rye (*Leymus condensatus*) are popular in larger settings, natural landscaping, and native plant gardens. There are *Miscanthus* grasses whose variegations are horizontal, and appear even on a cloudy day to be stippled with sunshine. Many *Miscanthus* and *Pennisetum* species flower in mid or late summer, and the seed heads are long lasting, often remaining well into the winter. Some *Stipa* species flower in the spring, the inflorescence standing almost two metres above the clumps of leaves, and again the seed heads last late into the winter.

When gardening near natural wildland-urban interfaces, care to avoid invasive species, such as *Cortaderia jubata*, *Pennisetum setaceum*, and *Nassella tenuissima* (syn. *Stipa tenuissima*), is responsible horticulture.

Perennial Plant

Common chicory, *Cichorium intybus*, is a herbaceous perennial plant

A perennial plant or simply perennial is a plant that lives for more than two years. The term (*per-* + *-ennial*, "through the years") is often used to differentiate a plant from shorter-lived annuals and biennials. The term is also widely used to distinguish plants with little or no woody growth from trees and shrubs, which are also technically perennials.

Perennials, especially small flowering plants, that grow and bloom over the spring and summer, die back every autumn and winter, and then return in the spring from their rootstock, are known as herbaceous perennials. However, depending on the rigors of local climate, a plant that is a perennial in its native habitat, or in a milder garden, may be treated by a gardener as an annual and planted out every year, from seed, from cuttings or from divisions. Tomato vines, for example, live several years in their natural tropical/subtropical habitat but are grown as annuals in temperate regions because they don't survive the winter.

There is also a class of evergreen, or non-herbaceous, perennials, including plants like *Bergenia* which retain a mantle of leaves throughout the year. An intermediate class of plants is known as subshrubs, which retain a vestigial woody structure in winter, e.g. *Penstemon*. The local climate may dictate whether plants are treated as shrubs or perennials. For instance, many varieties of *Fuchsia* are shrubs in warm regions, but in colder temperate climates may be cut to the ground every year as a result of winter frosts.

The symbol for a perennial plant, based on *Species Plantarum* by Linnaeus, is ♃, which is also the astronomical symbol for the planet Jupiter.

Life Cycle and Structure

Perennial plants can be short-lived (only a few years) or they can be long-lived, as are some woody plants like trees. They include a wide assortment of plant groups from ferns and liverworts to the highly diverse flowering plants like orchids and grasses.

Plants that flower and fruit only once and then die are termed monocarpic or semelparous. However, most perennials are polycarpic (or iteroparous), flowering over many seasons in their lifetime.

Perennials typically grow structures that allow them to adapt to living from one year to the next through a form of vegetative reproduction rather than seeding. These structures include bulbs, tubers, woody crowns, rhizomes plus others. They might have specialized stems or crowns that allow them to survive periods of dormancy over cold or dry seasons during the year. Annuals produce seeds to continue the species as a new generation while the growing season is suitable, and the seeds survive over the cold or dry period to begin growth when the conditions are again suitable.

Many perennials have developed specialized features that allow them to survive extreme climatic and environmental conditions. Some have adapted to survive hot and dry conditions or cold temperatures. Those plants tend to invest a lot of resource into their adaptations and often do not flower and set seed until after a few years of growth. Many perennials produce relatively large seeds, which can have an advantage, with larger seedlings produced after germination that can better compete with other plants. Some annuals produce many more seeds per plant in one season, while some (polycarpic) perennials are not under the same pressure to produce large numbers of seeds but can produce seeds over many years.

Dahlia plants are perennial.

Growth

In warmer and more favorable climates, perennials grow continuously. In seasonal climates, their growth is limited to the growing season.

In some species, perennials retain their foliage all year round; these are evergreen perennials. Other plants are deciduous perennials, for example, in temperate regions a perennial plant may grow and bloom during the warm part of the year, with the foliage dying back in the winter. In many parts of the world, seasonality is expressed as wet and dry periods rather than warm and cold periods, and deciduous perennials lose their leaves in the dry season.

With their roots protected below ground in the soil layer, perennial plants are notably tolerant of wildfire. Herbaceous perennials are also able to tolerate the extremes of cold in temperate and Arctic winters, with less sensitivity than trees or shrubs.

Benefits in Agriculture

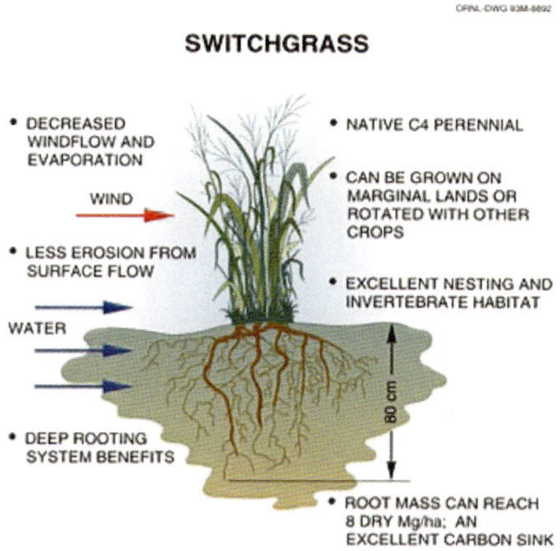

Switchgrass is a deep-rooted perennial. These roots are more than 3 meters long.

Although most of humanity is fed by the re-sowing of the seeds of annual grain crops, (either naturally or by the manual efforts of man), perennial crops provide numerous benefits. Perennial plants often have deep, extensive root systems which can hold soil to prevent erosion, capture dissolved nitrogen before it can contaminate ground and surface water, and out-compete weeds (reducing the need for herbicides). These potential benefits of perennials have resulted in new attempts to increase the seed yield of perennial species, which could result in the creation of new perennial grain crops. Some examples of new perennial crops being developed are perennial rice and intermediate wheatgrass. The Land Institute estimates that profitable, productive perennial grain crops will take at least 25 years to achieve.

Location

Perennial plants dominate many natural ecosystems on land and in fresh water, with only a very few (e.g. *Zostera*) occurring in shallow sea water. Herbaceous perennial plants are particularly dominant in conditions too fire-prone for trees and shrubs, e.g., most plants on prairies and steppes are perennials; they are also dominant on tundra too cold for tree growth. Nearly all forest plants are perennials, including the trees and shrubs.

Perennial plants are usually better competitors than annual plants, especially under stable, resource-poor conditions. This is due to the development of larger root systems which can access water and soil nutrients deeper in the soil and to earlier emergence in the spring.

Types

- Examples of evergreen perennials include *Begonia* and banana.
- Examples of deciduous perennials include goldenrod and mint.
- Examples of monocarpic perennials include *Agave* and some species of *Streptocarpus*.
- Examples of woody perennials include maple, pine, and apple trees.
- Examples of herbaceous perennials used in agriculture include alfalfa, *Thinopyrum intermedium*, and Red clover.

Houseplant

A houseplant is a plant that is grown indoors in places such as residences and offices. Houseplants are commonly grown for decorative purposes, but studies have also shown them to have positive psychological effects. Houseplants also help with indoor air purification. Plants used in this fashion are most commonly, though not always, tropical or semi-tropical epiphytes, succulents or cacti.

Houseplants need the correct moisture, light levels, soil mixture, temperature, and humidity. As well, houseplants need the proper fertilizer and correct-sized pots.

Plant Requirements

Major factors that should be considered when caring for houseplants are moisture, light, soil mixture, temperature, humidity, fertilizers, potting, and pest control. The following includes some general guidelines for houseplant care. For specific houseplant needs, the tags that sometimes come with plants are notoriously unhelpful and generic. Specific care information may be found widely online and in books.

Moisture

Both under-watering and over-watering can be detrimental to a houseplant; the soil needs to be moist but not flooded.

Light

Different plants require different amounts of light, for different durations. Photoperiodism must also be considered, since some plants such as Poinsettia and Schlumbergera are influenced by either decreasing or increasing daylight hours.

Most houseplants must rely on window light. It is possible to supplement this with artificial lighting of suitable wavelengths.

Soil

Houseplants are generally grown in specialized soils called *potting compost* or *potting soil*, not in local natural soil. A good potting compost mixture includes soil conditioners to provide the plant with nutrients, support, adequate drainage, and proper aeration. Most potting composts contain a combination of peat and vermiculite or perlite. Concern over environmental damage to peat bogs, however, is leading to the replacement of peat by coir (coconut fibre), which is a sustainable resource. Sterilised soil can also be used.

Temperature

Most houseplants are tropical species selected for their adaptation to growth in a climate which ranges from 15 °C to 25 °C (60 °F to 80 °F), similar to the temperature in most homes. Temperature control for other plants with differing requirements needs attention to heating and/or cooling.

Humidity

Humidity is slightly more difficult to control than temperature. The more commonly used houseplants have established that they can survive in low humidity environments as long as their roots are kept properly irrigated. Most plants thrive in 80% relative humidity while most homes are usually kept around 20% to 60% relative humidity.

Fertilizers

Plants require soil minerals, mainly nitrate, phosphate, and potassium. Nitrogen is essential for green, leafy growth. Phosphorus is essential for flowering or fruiting plants. Potassium is essential for strong roots and increased nutrient uptake. Minor and trace elements, such as calcium, magnesium and iron, may also be necessary.

Pot Types and Sizes

Proper pot size is an important factor to consider. A pot that is too large will cause root disease because of the excess moisture retained in the soil, while a pot that is too small will restrict a plant's growth. Generally, a plant can stay in the same pot for two or so years. Pots come in a variety of types as well, but usually can be broken down into two groups: porous and non-porous. Porous pots are usually clay and are highly recommended because they provide better aeration as air passes laterally through the sides of the pot. Non-porous pots such as glazed or plastic pots tend to hold moisture longer and restrict airflow. Another needed feature is drainage holes. Usually pots come with holes in the bottom to allow excess water to flow out of the soil which helps to prevent root rot. If a pot does not have drainage holes, it is best to double pot that plant so the inner pot can be lifted out and the excess water accumulated in the bottom of the outer pot can be removed. Soak old pots thoroughly in a solution of 1 part bleach to 10 parts water to kill any bacteria that may remain.

Effect on Indoor Air Pollution

Indoor plants reduce components of indoor air pollution, particularly volatile organic compounds (VOC) such as benzene, toluene, and xylene. The compounds are removed primarily by soil microorganisms. Plants can also remove CO_2 "Carbon dioxide", which is correlated with lower work performance, from indoor areas. The effect has been investigated by NASA for use in spacecraft. Plants also appear to reduce airborne microbes and increase humidity.

Alternative Growing Methods

Hydroponics

Aside from traditional soil mixtures, media such as expanded clay may be employed in hydroponics, in which the plant is grown in a water and nutrient solution. This technique has a number of benefits, including an odorless, reusable, and more hygienic media. Any habitat for soil-bound pests is also eliminated, and the plant's water supply is less variable. However, some plants do not grow well with this technique, and media is often difficult to find in some parts of the world, such as North America, where hydroponics and specifically hydroculture is not as well-known or widespread.

Subirrigation

Subirrigation offers another alternative to top-watering techniques. In this approach the plant is watered from the bottom of the pot. Water is transferred up into the potting media (be it soil or others) by capillary action. Advantages of this technique include controlled amounts of water, resulting in lower chances of overwatering if done correctly, no need to drain plants after watering unlike traditional top-water methods, and less compaction of the media due to the pressure put on the media from top-watering.

Cut Flowers

Rose, hydrangea and calla wedding bouquet

Garland sellers outside Banke Bihari Temple, Vrindavan, India

A flower market in Vietnam

Flower display in a US supermarket

Cut flowers are flowers or flower buds (often with some stem and leaf) that have been cut from the plant bearing it. It is usually removed from the plant for decorative use. Typical uses are in vase displays, wreaths and garlands. Many gardeners harvest their own cut flowers from domestic gardens, but there is a significant floral industry for cut flowers in most countries. The plants cropped vary by climate, culture and the level of wealth locally. Often the plants are raised specifically for the purpose, in field or glasshouse growing conditions. Cut flowers can also be harvested from the wild.

The cultivation and trade of flowers is a specialization in horticulture, specifically floriculture.

Uses

A common use is for floristry, usually for decoration inside a house or building. Typically the cut flowers are placed in a vase. A number of similar types of decorations are used, especially in larger buildings and at events such as weddings. These are often decorated with additional foliage. In some cultures, a major use of cut flowers is for worship; this can be seen especially in south and southeast Asia.

Sometimes the flowers are picked rather than cut, without any significant leaf or stem. Such flowers may be used for wearing in hair, or in a button-hole. Masses of flowers may be used for sprinkling, in a similar way to confetti.

Garlands, wreaths and bouquets are major derived and value added products.

Longevity

Once flowers are removed from the plant they continue to grow slowly, but have a diminished capability of receiving the nutrients that are vital for their survival. In most countries, cut flowers are a local crop because of their perishable nature. In India, much of the product has a shelf life of only a day. Among these are marigold flowers for garlands and temples, which are typically harvested before dawn, and discarded after use the same day.

The majority of cut flowers can be expected to last several days with proper care. This generally requires standing them in water in shade. They can be treated in various ways to increase their life. According to James C. Schmidt, a horticulturist at the University of Illinois, originally putting cut flowers in a sterilized vase is important to extending the life of the flowers. Vases can be cleaned using a household dish detergent or a combination of water and bleach. Using these disinfectants ensures that there will be less bacteria growing within the vase that could potentially cause the plant to wilt and die at a faster rate. Schmidt also claims that cutting the flowers diagonally with a sharp knife under running water ensures that they can immediately take up fresh and clean water. Re-cutting the stems periodically will ensure that there is a fresh surface from which the stems can take up water. This will allow the flowers to last even longer. Other ways to care for vase flowers includes keeping flowers away from ceiling fans and air-conditioning vents as this can lead to dehydration, keeping flowers away from fresh fruit of vegetables, using filtered water rather than tap water so as to avoid both chlorine and fluoride, and keeping flowers away from your television.

There is also a market for 'everlasting' or dried flowers, which include species such as Xerochrysum bracteatum. These can have a very long shelf life.

Additives

According to the Brooklyn Botanical Garden, different additives can be used to prolong the lives of fresh cut flowers. Experiments were performed with various substances mixed with water, including aspirin, vitamin pills, vinegar, pennies, and flower food to test their effect on cut flowers' lifespans. Each plant was placed in the same environment and the same type of plant was used in each vase.

This research found that the best additive for flowers was the retailer-provided "flower food" that is usually given with a bouquet. Plants are known to thrive in an environment where there are few bacteria, plenty of food for energy, and water uptake is encouraged. Flower foods contain an acidifier that helps to adjust the water's pH. With a lower pH the water and food conducting system within the flower can work at maximum efficiency. The sugar in the food will be used by the plant as an energy source, which had been lost when the flower was cut away from its root. With these nutrients the plant will be able to fully develop. Finally, there are stem unpluggers that will make sure that the flower can easily take up water and nutrients that can later be used to take care of the needs of the rest of the plant. This combination gives the fresh cut flowers everything that they need to survive longer. When tests were carried out in St. Mary's College C.S.S.p, Rathmines, Dublin, however, results showed that glucose was more effective at prolonging the life of cut flowers than the commercial plant food. The experiment was carried out as part of the Junior Certificate Science Examination set by the Irish State Examinations Commission 2008.

Trade

The largest producers are, in order of cultivated area, China, India, and the United States. The largest importer and exporter by value is the Netherlands, which is both a grower and a redistributor of crops imported from other countries. Most of its exports go to its European neighbours.

In recent decades, with the increasing use of air freight, it has become economic for high value crops to be grown far from their point of sale; the market is usually in industrialised countries. Typical of these is the production of roses in Ecuador and Colombia, mainly for the US market, and production in Kenya and Uganda for the European market. Some countries specialise in especially high value products, such as orchids from Singapore and Thailand.

The total market value in most countries is considerable. It has been estimated at approximately GBP 2 billion in the United Kingdom, of the same order as that of music sales.

Floristry

Floristry is the production, commerce and trade in flowers. It encompasses flower care and handling, floral design, or flower arranging, merchandising, and display and flower delivery. Wholesale florists sell bulk flowers and related supplies to professionals in the trade. Retail florists offer fresh flowers and related products and services to consumers. The first flower shop opened in 1875.

A Parisian Flower Market by Victor Gabriel Gilbert

A wedding bouquet of cymbidium arranged by a florist

Floristry can involve the cultivation of flowers as well as their arrangement, and to the business of selling them. Much of the raw material supplied for the floristry trade comes from the cut flowers industry. Florist shops, along with online stores, are the main flower-only outlets, but supermarkets, garden supply stores, and filling stations also sell flowers.

Floral design or floral arts is the art of creating flower arrangements in vases, bowls, baskets, or other containers, or making bouquets and compositions from cut flowers, foliages, herbs, ornamental grasses, and other plant materials. Often the terms "floral design" and "floristry" are considered synonymous. Florists are people who work with flowers and plants, generally at the retail level. Floristry differs from floristics, the study of distribution and relationships of plant species over geographic areas. Floristry also differs from horticulture, which more broadly relates to the cultivation of flowers and plants so they will remain fresh as long as possible, and would be desirable for purchase, which also involves knowledge of customers' requirements and expectations. The ability to create a variety of floral designs such as wreaths, bouquets, corsages, boutonnières/'buttonholes', permanent arrangements, and other more complicated arrangements are also important.

Education, both formal and informal, is another significant segment of the floristry industry. Established floristry designers and artists impart their craft to students interested in floral design as hobby or career. Courses are generally available through community colleges, private post-secondary vocational schools, and professional florist trade associations.

In the Netherlands, the first horticultural college was founded in 1896 in Naaldwijk; the second horticultural college was founded in 1897 in Aalsmeer. In 1926, the first national professional qualification examinations in floristry were held in the Netherlands. The horticultural college in Aalsmeer celebrated its 75th anniversary in 1972 and in 1997 its 100th anniversary. Since 1926 is the horticultural college in Aalsmeer was called the 'Rijks Middelbare Tuinbouw School' (RTMS). The first professional floristry education started at the RMTS in 1968. The first professor in floristry at the horticultural college in Aalsmeer was Mr. Theo Boerma. In 1972 Theo Boerma started teaching professional evening courses for the floristry diploma: vakdiploma Bloemist-Winkelier. In 1980 Theo Boerma and his wife José Boerma founded the first privately owned floristry school; Boerma Instituut International Floral Design School in Aalsmeer. The floristry diploma was recognized by the Dutch government until 1996. When the borders of Europe opened, the diploma was no longer needed but professional training for Dutch and international students is still organised by the Boerma Instituut.

The floristry business has a significant market in the corporate and social event world, as flowers play a large part in the decor of special events and meetings. Centerpieces, entryways, reception tables, bridal bouquets, wedding chuppahs, and stage sets are only a few examples of how flowers are used in the business and social event settings. Flowers are also traditionally used in ecclesiastical settings and their arrangement is often done by skilled church volunteers.

Styles

Many nations have their own style of floral arranging. This is dependent on what flower varieties are readily available, and the culture of the nation.

Ikebana

Child learning the art of Ikebana. Today, florists in Japan also sell pre-arranged or done-while-you-wait ikebanas to go.

Ikebana is a style of floral design that originated in Japan. Best known for its simplicity of line and form, Ikebana is a design style primarily practiced for personal enjoyment. It has three parts of alignment: heaven, man, and Earth.

English Garden

English Garden style is traditionally an English form of floral design. Stems are placed in a radial fashion and feature abundant use of seasonal flowers and foliages. These designs are often done as low, tufted mounds, or taller vase arrangements that are all-sided (360°), and incorporate garden flowers like roses, gardenia, camellia, delphinium and peonies. The flowers are usually arranged with minimal space between the blooms and foliage is used to accent the flowers as these are the main feature.

Modern/European Style

Modern/European Style floristry involves contemporary, linear designs that highlight unique forms of both individual floral materials and of the designs themselves. Arrangements generally feature negative space and incorporate asymmetric placement of materials. The style stands in direct contrast to traditional radial arrangements such as English Garden. Modern designs are identified by their play on the space used between each bloom, which is often dramatic, and the play on the use of color and different textures, which can be quite experimental.

Modern/European Style designs often incorporate unique, exotic or tropical flowers such as Bird of Paradise, Orchids, Anthurium, and Protea of the Caribbean and other exotic locals, but may also employ more common blossoms, such as Roses, Gerberas and lilies.

Modern/Dutch Style

Contemporary/Dutch Style designs also involve contemporary and linear designs. The Dutch designs use a lot of different kinds of greens are used in a natural way. The “Dutch Garden” style arrangement -started in the early 80’s- is a very good example of a Dutch style arrangement. Stones, bark and mosses are used in these designs.

Significance of Particular Flowers

Flower market in Aalsmeer

Tulip shop in Amsterdam

Typically, a florist will organize flowers by season and holiday. Flowers have various different meanings in different cultures. The holidays and events for which flowers are used vary. Poppies are used to remember fallen soldiers only in Great Britain and the Commonwealth countries. People often prefer flowers that are associated with their ethnic group or country and the cultural meaning of the flower color strongly affects their choice and use. Colors convey different meanings to different groups so that, for example, various colors may be associated with luck or death or love. A flower such as a red rose might communicate love in some cultures, whereas in others, it might be considered indecent or simply puzzling. The vastly divergent attitudes to the color white can also lead to miscommunication and sometimes give offense. White, for example, represents death in many Asian cultures, whereas white is usually a symbol of purity and innocence in countries with a European heritage.

Fresh Flower Sources

There are usually three major sources of fresh flowers for retail florists: local growers, local wholesalers and flower auctions. The majority of these are supplied by specialist horticultural and cut flower suppliers.

Internationally there are hundreds of wholesale flower markets and auctions, the largest of which is located in Aalsmeer, the Netherlands: the Bloemenveiling Aalsmeer. Other major markets include the fledgling Dubai Flower Centre and the Ota Flower Market in Tokyo, Japan.

Within North America, Miami is considered to be the main distribution point for imported fresh flowers. Many local wholesalers purchase fresh flower stock from importers in Miami to resell to local florists in their areas. Wholesale flower districts are present in many North American cities such as New York, Boston, and Los Angeles, San Francisco and Carlsbad, California. Flower auctions are run using the Dutch clock system and are mainly located in Canada: Toronto, Montreal, and Vancouver.

Floristry Shops

In the Netherlands gardners start to sell flower arrangements from around 1850. The first 'real' flower shops started to open around 1900 in the Netherlands. At that time the flower shops also had small greenhouses where they sold their flowers from.

A name tribute ("MUM") at a funeral in England, made by a florist using chrysanthemums

Generally, a florist's shop will contain a large array of flowers, sometimes displayed on the street, or will have a large plate glass window to display the flowers. To keep them fresh, the flowers will be refrigerated and kept in water, generally in glass or plastic vases or other containers. Most shops have a refrigerator/chiller cabinet/ cooler near the front of the shop/store with large glass doors so that customers can easily view the contents. Some shops also have another cooler out of the customers' view where they keep extra stock and arrangements for customers' orders. Most stores have a separate work room in which the design florist or designers can work on orders with more privacy.

The flowers sold in florist shops typically represent the varieties abundantly available in the season but also include blossoms flown in from around the world. Basic varieties in temperate latitudes include roses, tulips, irises, orchids and lilies. Fashion sometimes plays a role in floristry; what is considered the flower that everyone needs to have today can change very quickly.

Some shops also stock gift baskets, fruits, and chocolates as well as flowers, whereas some shops will purchase these things only when needed for an order. Floral business is seasonal and is heavily influenced by the following holidays and events: Christmas, Valentine's Day, Administrative Professionals' Day, Mothers' Day, All Souls Day, Advent, Easter, weddings and funerals. These occasions make up the largest part of the business, with the sale of house plants and home decor being a smaller, but more constant, part. Flowers for personal enjoyment as well as those selected to celebrate birthdays, anniversaries, thank-yous and get well wishes are also a significant portion of a florist's business.

Supermarket flowers

Supermarket display under fluorescent lighting

Cut flowers are widely sold in supermarkets and petrol stations. These outlets offer a limited selection, usually in the form of mixed bouquets and roses by the dozen. Flowers purchased at these outlets are generally less expensive than the flowers purchased at a florist shop. Professional florists may offer a greater selection, higher flower quality, or more skillfully arranged flowers.

Technology

In the United States, the number of florists has dropped since a peak of over 27,000 in 1992 to fewer than 20,000 in 2007. The World Wide Web has contributed to this decline by making alternatives more accessible.

Floral wire services play an intermediary by gathering orders to be fulfilled by local florists. Floral wire services take a percentage of customers order value for orders provided through their websites and additional fees for the transmission of orders to local florists.

Few companies ship flowers direct to the consumer. ProFlowers, Interflora, Teleflora, FTD typically function as *order gatherers* and will work with local florists to have the order delivered.

However, younger companies such as The Bouqs are building relationships and exclusive deals with flower farms to enable consumers to receive flowers directly from the farm.

Floral Industry

The floral industry is one of the higher industries in many developing and underdeveloped countries. Floriculture as an industry began in the United Kingdom, where flowers were grown on a large scale on the vast estates. The present day floral industry is a dynamic, global, fast-growing industry, which has achieved significant growth rates during the past few decades. In the 1950s, the global flower trade was less than US$3 billion. By 1994, it had grown to US$100 billion. In recent years, the floral industry has grown six percent annually, while the global trade volume in 2003 was US$101.84 billion.

The floral industry essentially consists of three major components: the growers, the wholesalers and the retailers whose businesses are quite intermingled. The recent trends are more towards eliminating the intermediaries, the wholesalers between the growers and the retailers, so that the flowers are made available at considerably low prices.

Transportation

Some flowers are sent packed flat in boxes. This enables large amounts of flowers to be packed in small spaces like aircraft holds. Other flowers cannot survive for long periods out of water such as orchids, gerberas (gerber daisies) and water lilies. These are either sent with their own sealed water container (called picks) on each stem end - for more expensive or tropical flowers - or are transported in buckets of water (This method of transport in water is often referred to as "Procona"). The latter method extends the life of flowers and reduces labor time as flowers are ready for sale, but obviously also reduces the amount of flowers that can be transported as they are much heavier than dry-packed flowers and hence air transportation charges are higher.

Flowers take a number of routes to the consumer, depending on where they are grown and how they are to be sold. Some growers cut and pack flowers at their nurseries, sending them directly out to the consumer by mail order. Some flowers are sent to packing companies, who grade the flowers and arrange them in bunches for sale to supermarkets or to deliver by mail order. Some flowers are graded and sleeved by the growers and sold at wholesale flower markets; the wholesalers then sell them on to florists who condition and arrange the flowers for the consumer.

The Netherlands and the History of the Flower Industry

Traditionally, the center of flower production has been near their largest consumers: the developed world, where Japan, Western Europe and North America were both major producers and consumers. The major consumer markets being Germany (22 percent), the United States (15 percent), France (10 percent), the United Kingdom (10 percent), the Netherlands (9 percent), Japan (6 percent), Italy (5 percent), and Switzerland (5 percent).

The Netherlands remains the center of production for the European floral market, as well as a major international supplier to other continents. The flower auction at Aalsmeer is the largest flower market in the world. Since the mid-1970s, the production and distribution of cut flowers in Netherlands has burgeoned. In 1995, Dutch growers produced over 8 billion blooms and the flower auctions collectively traded more than 5.4 billion guilders (about $3.2 billion) in cut flowers and potted plants, contributing over 4 billion guilders annually to the Dutch balance of trade.

New Flower Growing Centres

Experts believe that the production focus has moved from traditional growers to countries where the climates are better and production and labor costs are lower. This has resulted in a paradigm shift in the floral industry. The Netherlands, for instance, has already shifted attention from flower production to flower trading, though it plays an important role still in the development of floricultural genetics. The new centers of production are typically developing countries like Ecuador (The Biggest producer and export of roses worldwide), Colombia (second largest exporter in the world and with a market of more than 40 years old), Ethiopia, Kenya, and India. Other players in this global industry are Israel, South Africa, Australia, Thailand and Malaysia. New Zealand, due to its position in the Southern Hemisphere, is a common source for seasonal flowers that are typically unavailable in Europe and North America.

In Africa, Kenya is the largest exporter, supplying a large percentage of Europe's flowers, the industry there is represented by the Kenya Flower Council.

In South America, Colombia is the leading flower producer and exporter accounting for 59% of all flowers imported to The United States in 2006. The United States imports 82% of its flowers. Growers in the United States state that 3 out of 4 flowers in the United States are grown outside the US with Colombia being the biggest exporter. The United States signed a free trade agreement with Colombia and that has lowered the cost of Colombian flowers in the United States. Ecuador has become, in recent years, the leading South American rose producer and is well known throughout the world for its high quality, large headed roses due to the high altitude location of its rose farms.

Companies

- Florists' Transworld Delivery (FTD) (USA)
- Teleflora (USA)
- 1800 Flowers (USA)
- Euroflorist (Sweden)
- Tokyoflora.com (Japan)
- Dutch Flower Group (The Netherlands)
- HilverdadeBoer (The Netherlands)
- FleuraMetz (The Netherlands)

Trade Organisations

- Royal FloraHolland
- Society of American Florists
- British Florist Association

Floral Design

Floral art in a competition at the Chandigarh Rose Festival

Floral design or flower arrangement is the art of using plant materials and flowers to create a pleasing and balanced composition. Evidence of refined floristry is found as far back as the culture of ancient Egypt. Professionally designed floral designs, arrangements or artwork incorporate the elements of floral design: line, form, space, texture, and color, and the principles of floral design: balance, proportion, rhythm, contrast, harmony, and unity.

There are many styles of floral design including Botanical Style, Garden Style, Crescent Corsage, Nosegay Corsage, Pot et Fleur, Inverted "T", Parallel Systems, Western Line, Hedgerow Design, Mille de Fleur, and Formal Linear to name a few.

The Eastern, Western, and European styles have all influenced the commercial floral industry as it is today. Ikebana is a Japanese or eastern style of floral design, and incorporates the three main line placements of heaven, man and earth. In contrast, the European style emphasizes color and variety of botanical materials not limited to just blooming flowers, in mass gatherings of multiple flowers. Western design historically is characterized by symmetrical, asymmetrical, horizontal, and vertical style of arrangements.

In addition to flower arrangements, floral design includes making wreaths, nosegays, garlands, festoons, boutonnieres, corsages, and bows.

Dried Arrangements and Related Media

Permanent creations and components incorporating dried materials such as bark, wood, dried flowers, dried (and often aromatic) inflorescences, leaves, leaf skeletons, preserved materials and artefacts, are common extensions of the art floral design, and are of practical importance in that they last indefinitely and are independent of the seasons. Their materials offer effects, idioms, and associations complementary to, and contrasting with, fresh flowers and foliage.

Floral designers at work in Germany

Floral design

A small design

Chrysanthemum arrangement for catering

Schools

With the ever-growing interest in the natural world and flowers, the floral industry continues to grow. The increase in educational institutes providing training in floral design has expanded to many state universities as well as certified design schools worldwide.

Associations

Prominent industry associations that promote floral design worldwide include the American Institute of Floral Designers (AFID), the Society of American Florists (SAF), and the National Association of Flower Arranging Societies (NAFAS). These and other associations promote floral design through workshops, conferences, flower shows, and seminars.

Designers

Chrysanthemums in an antique bottle. Floral design is commonly an element of editorial photography.

Notable floral designers include Daniel Ost, Junichi Kakizaki, Paula Pryke, Phil Rulloda, Catherine Conlin, Constance Spry, Jennifer McGarigle, Judith Blacklock, Stanlee Gatti, Irene Hayes, Julia Clements, Azuma Makoto, and the White House Chief Floral Designer.

Flower Preservation

Preserved rose blossoms and silk flowers

Flower preservation is as early as the history of man, although deliberate flower preservation is a more recent phenomenon. In the Middle East, the bones of pre-historic man were discovered with delicate wild flowers probably as a tribute to a passing loved one.

Evidence of deliberate use of specific flowers is indicated by the pollen grains that were present. Brightly colored and vivid flowers were also found in Egyptian tombs. These flowers were approximated to be 4,000 years old. Medicinal nosegays began to give way to ornamental ones. Flowers essentially started to be used for decorative purposes such as jewels, fans and gloves. During the Elizabethan Age the once familiar ruff was replaced by soft lacy collars, and bosom flowers also became popular.

Out of the Victorian era grew the fascination of communicating with flowers carried in the nosegays. The idea of the Language of Flowers developed, when it was decided that giving and receiving a bouquet of flowers, when the flowers themselves carry a meaning, gives much greater pleasure.

Drying

One of the earliest methods of preserving flowers is by drying. Many plants retain their shape and color when air-dried naturally.

Use of glycerine, making the preserved plant supple and long-lasting. To use this method, the plant material needs to be gathered in a fully hydrated state. Water and glycerine are then mixed. The ratio of water to glycerine should be 2:1. The water should be lukewarm for better mixing and faster absorption. If the autumn colors are showing, it may be too late to preserve them in glycerine.

Pressing is a very easy way to preserve flowers although the relief is lost and the flowers are flat. Unglazed paper, such as newsprint or an old telephone book, is best for pressing. Flowers are spread so they do not overlap between several thicknesses of newspaper. Additional layers of paper and flowers can be built up and then covered with a board or piece of cardboard before pressing down with a heavy object. The time required for drying, depending on the flower size or tissue content, can be anywhere from two to four weeks.

Modern Preservation Methods

Hanging Flowers in Ventilated Area

There are several air-drying methods used for different kinds of flowers. The easiest and most effective way to dry most flowers is to tie them in small bundles with twine, raffia or ribbons, and hang them upside down, out of direct sunlight in a warm, well-ventilated place. Bunches should contain one type of flower. Large flowers are dried individually. The leaves stripped as soon as possible after picking. They retain moisture and slow down the drying process. Wall hooks, poles or wires are placed at least six inches from the ceiling. Heads should be staggered to allow plenty of air to circulate. This prevents mildew and rot. It may be necessary to re-tie bunches or individual flowers half way through the drying process, because the stems tend to shrink as they dry.

The drying period can range from one week to several weeks depending on the type of material, when and where it was harvested and the humidity of the place where it is drying. The stems of hung flowers tend to dry unnaturally straight and the flowers become very brittle.

Suggested flowers to air dry include: strawflowers, goldenrod, hydrangeas, celosia (crested and plumed types), globe amaranth, salvia, Xeranthemum and many of the "everlastings."

Hot Air Drying in Drying Chamber

Compact flowers like marigolds, chrysanthemums, cornflowers and zinnias dry well in a fan-assisted, convection chamber. Non-ventilated chambers are not appropriate, because they generate too much moisture. The material must be dried at a low temperature (30-35°Celsius), over many hours. The flowers are slotted through holes in a wire mesh rack leaving room for the stems to dangle below. The time required depends upon the density of the flowers.

Silica Gel Drying

Another trend is silica gel. Its initial cost is greater than that of borax-sand or borax-cornmeal combinations, but silica gel can be used over and over for many years. Silica gel dries flowers quickly, so it can be used to dry more flowers during a single season than the same quantity of a borax mixture. Silica gel is available under a number of trade names. It is white, but some types contain blue crystals that act as an indicator of the amount of moisture that has been absorbed. When these crystals are clear blue, the material is dry. As moisture is absorbed from the flowers, the crystals gradually turn pink. At that point, it is time to re-dry the crystals before using them again.

To dry the material, silica gel is spread on open pans or cookie sheets in a layer ½ to ¾ inch thick. Materials to be preserved are then baked in an oven at 250°Celsius for about an hour, or until the moisture-indicating crystals, if present, are blue again. Material is then stirred several times while drying.

Flowers dried in silica gel must be placed in airtight containers. If a container is not sealed tightly, the silica gel absorbs moisture from the air, and flowers dry too slowly or not at all. A candy tin, plastic container, coffee can, large-mouth jar or any other container with a tight-fitting lid may be used. If no containers with tight lids are available, loose tops should be sealed with tape. Silica gel is especially useful for drying fragile plants and flowers with delicate colors. Flowers that dry best in silica gel are allium, anemone, cornflower, roses, tulip and zinnia.

Molecular Sieve

Molecular sieve is a material containing tiny pores of a precise and uniform size that is used as an absorbent for gases and liquids. They are metal alumino silicates which have a crystalline structure consisting of an assembly of tetrahedral. The tetrahedral are made up of 4 oxygen atoms which occupy the summits surrounding either 1 silicon atom or an aluminium atom placed in the center. Compensating cations (sodium, potassium) make the hole electrically neutral. The hole forms an assembly of small cells (or pores) of uniform and known size, in which a molecule of smaller size can be trapped by the phenomenon known as absorption.

To ensure the dehydration of the fresh natural flowers the mixture of organic solvents is poured onto the hole until the level exceeds the level of the flowers by about 2 cm. The water molecules are progressively absorbed into the small cells or pores of the molecular sieve.

The receptacle is closed hermetically for a few days.

Once dried, the flower reabsorbs a little moisture of atmospheric origin, and this increases its suppleness and its plasticity.

As molecular sieve relates to the long-term preservation and treatment of cut flowers for long duration, (i.e. continually maintaining their decorative properties), Vermont Flowers has been using this method of preserving flowers instead of the traditional silica gel more than 20 years ago. The advantage was a much more professional regeneration of the molecular sieves. Flowers which are particularly suitable for such a treatment are roses, peonies, camellias, marigolds, globe flowers, orchids, dahlias, carnations, phloxes, summer chrysanthemums, hollyhocks, and the like, and other species with many petals or a fairly rigid structure.

Freeze Drying

Originally introduced in 1813 by William Hyde Wollaston to the Royal Society in London, it was not until the late 80's the freeze-drying industry discovered the allurement and longevity of freeze-dried flowers.

Freeze dried flowers are fresh flowers that have been specially dried to preserve their natural shape and color. Freeze drying is accomplished by a process called sublimation. It requires a special freeze-drying machine.

It involves first freezing the flowers at 100K for at least 12 hours. A vacuum pump slowly pulls the moisture out of the flowers as a vapor in one chamber, and then the vapor condenses as ice in another chamber. Because of this process, the shape and natural color of the flower is maintained. It has been found that certain flowers retain their color well despite the fact they have been freeze-dried. Apparently, such flowers retain their color due to the tissue composition of the petals, leaves, and the like. Carnations, African violets, roses, asparagus and other ferns, and baby's breath exhibit good color retention notwithstanding the dehydration during the freeze-drying process. Those floral pieces which either dull or fade from dehydration may be given color by utilization of a florist's spray tint. This spray coloring restores the lost color which, in the sealed environment of the glass container of the final product, retains its given color along with the natural color of the other pieces.

Pressed

This floral preservation has its roots in Victorian times, but has also been brought into the current list of Floral Preservation because of new techniques of pressing, vacuum sealing and color enhancing. Pressed Floral Preservation involves pressing the flowers, using many different types of presses, depending on the flower type. The art is then framed in a flat frame, to hang on the wall. Vacuum sealing the art is recommended and the use of Museum Quality glass with a UV Factor of 97% to ensure the colors last.

References

- Scotland, R. W.; Wortley, A. H. (2003). "How many species of seed plants are there?". Taxon. 52 (1): 101–104. doi:10.2307/3647306. JSTOR 3647306
- Hitchmough, James; Fieldhouse, Ken (2004). Plant User Handbook. Wiley-Blackwell. pp. 292–309. ISBN 978-0-632-05843-3
- "Floriculture News" (PDF). No. 64. The Department of Agriculture, Western Australia. May 2005. Retrieved September 17, 2012
- Simonin, K. A.; Roddy, A. B. (2018). "Genome downsizing, physiological novelty, and the global dominance of flowering plants". PLOS Biology. 16 (1): e2003706. doi:10.1371/journal.pbio.2003706
- Fish, Martin (2004). Bedding plants (Collins practical gardener). United Kingdom: Collins. p. 160. ISBN 0007164068
- Smith, Ronald. "Methods of Preserving Flowers:Pressing". North Dakota State University publications. Retrieved 18 September 2011
- Soltis, Pamela S.; Soltis, Douglas E. (2004). "The origin and diversification of angiosperms". American Journal of Botany. 91 (10): 1614–1626. doi:10.3732/ajb.91.10.1614. PMID 21652312
- David Sadava; H. Craig Heller; Gordon H. Orians; William K. Purves; David M. Hillis (December 2006). Life: the science of biology. Macmillan. pp. 477–. ISBN 978-0-7167-7674-1. Retrieved 4 August 2010
- Soltis, D. E.; Soltis, P. S. (2004). "Amborella not a "basal angiosperm"? Not so fast". American Journal of Botany. 91 (6): 997–1001. doi:10.3732/ajb.91.6.997. PMID 21653455

- "As easy as APG III - Scientists revise the system of classifying flowering plants". The Linnean Society of London. 2009-10-08. Retrieved 2009-10-02
- Jiao, Yuannian; Wickett, No4rman J.; Ayyampalayam, Saravanaraj; Chanderbali, André S.; et al. (2011). "Ancestral polyploidy in seed plants and angiosperms". Nature. 473 (7345): 97–100. doi:10.1038/nature09916. PMID 21478875
- Kruchmol, Connie. "Revolutionary Wiggy Flowers". Bella Online--The Voice of women. Bella Online. Retrieved 18 May 2015
- Stewart, Wilson Nichols; Rothwell, Gar W. (1993). Paleobotany and the evolution of plants (2nd ed.). Cambridge Univ. Press. p. 498. ISBN 0-521-23315-1
- Wing, Xin; Duan, Shuying; Geng, Baoyin; Cui, Jinzhong; Yang, Yong (2007). "Schmeissneria: A missing link to angiosperms?". BMC Evolutionary Biology. 7: 14. doi:10.1186/1471-2148-7-14. PMC 1805421 . PMID 17284326
- The Garden Helper. The Difference Between Annual Plants and Perennial Plants in the Garden. Retrieved on 2008-06-22
- Thompson, Connie (5 May 2014). "Plenty of good local florists, but beware ordering flowers online". Retrieved 15 August 2014Keith Adams (December 2013). "Genomic Clues to the Ancestral Flowering Plant". Science. 342 (6165): 1456–1457. doi:10.1126/science.1248709. PMID 24357306
- "Processing of Horticultural Crops:Drying Flowers". FAO Corporate Document Repository. FAO. Retrieved 18 September 2011
- Raven, Peter H., Ray F. Evert, & Susan E. Eichhorn, 2005. Biology of Plants, 7th edition. (New York: W. H. Freeman and Company). ISBN 0-7167-1007-2
- Davies, T. J. (2004). "Darwin's abominable mystery: Insights from a supertree of the angiosperms". Proceedings of the National Academy of Sciences. 101 (7): 1904–9. doi:10.1073/pnas.0308127100. PMC 357025 . PMID 14766971
- Kolodny, Lora (25 June 2013). "Florist Seeded to Disrupt Blossom Trade With Farm-to-Home Delivery". Venture Capital Dispatch. Wall Street Journal. Retrieved 15 August 2014
- Bakker, Robert T. (17 August 1978). "Dinosaur Feeding Behaviour and the Origin of Flowering Plants". Nature. London: Macmillan. 274 (5672): 661–663. doi:10.1038/274661a

2

Flower Anatomy and Biology

A flower is the most crucial element of plant reproduction. This chapter discusses in detail the structure of a flower. It elucidates the core topics of plant reproductive morphology, inflorescence and flower development. It has been carefully written to provide an understanding of the varied aspects of the anatomy and biology of flowers.

Flower

A flower, sometimes known as a bloom or blossom, is the reproductive structure found in flowering plants (plants of the division Magnoliophyta, also called angiosperms). The biological function of a flower is to effect reproduction, usually by providing a mechanism for the union of sperm with eggs. Flowers may facilitate outcrossing (fusion of sperm and eggs from different individuals in a population) or allow selfing (fusion of sperm and egg from the same flower). Some flowers produce diaspores without fertilization (parthenocarpy). Flowers contain sporangia and are the site where gametophytes develop. Many flowers have evolved to be attractive to animals, so as to cause them to be vectors for the transfer of pollen. After fertilization, the ovary of the flower develops into fruit containing seeds.

In addition to facilitating the reproduction of flowering plants, flowers have long been admired and used by humans to bring beauty to their environment, and also as objects of romance, ritual, religion, medicine and as a source of food.

A poster with flowers or clusters of flowers produced by twelve species of flowering plants from different families.

Flowers in the Netherlands.

Morphology

Main parts of a mature flower (Ranunculus glaberrimus).

Diagram of flower parts.

Floral Parts

The essential parts of a flower can be considered in two parts: the vegetative part, consisting of petals and associated structures in the perianth, and the reproductive or sexual parts. A stereotypical flower consists of four kinds of structures attached to the tip of a short stalk. Each of these kinds of parts is arranged in a whorl on the receptacle. The four main whorls (starting from the base of the flower or lowest node and working upwards) are as follows:

Perianth

Collectively the calyx and corolla form the perianth:

- *Calyx*: the outermost whorl consisting of units called *sepals*; these are typically green and enclose the rest of the flower in the bud stage, however, they can be absent or prominent and petal-like in some species.
- *Corolla*: the next whorl toward the apex, composed of units called *petals*, which are typically thin, soft and colored to attract animals that help the process of pollination.

Reproductive

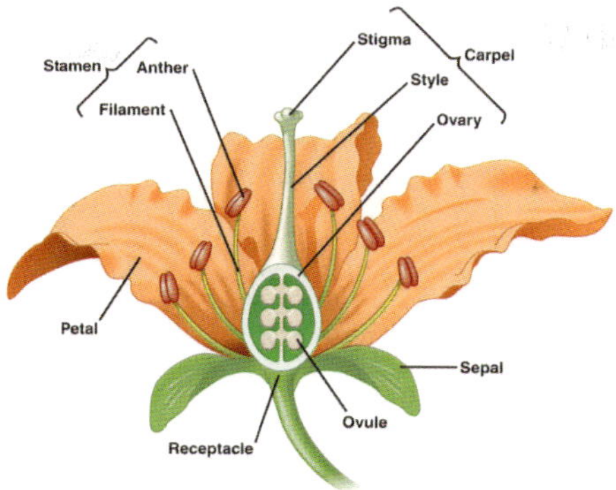

Reproductive parts of Easter Lily (*Lilium longiflorum*). 1. Stigma, 2. Style, 3. Stamens, 4. Filament, 5. Petal

- *Androecium* : the next whorl (sometimes multiplied into several whorls), consisting of units called stamens. Stamens consist of two parts: a stalk called a filament, topped by an anther where pollen is produced by meiosis and eventually dispersed.

- *Gynoecium* : the innermost whorl of a flower, consisting of one or more units called carpels. The carpel or multiple fused carpels form a hollow structure called an ovary, which produces ovules internally. Ovules are megasporangia and they in turn produce megaspores by meiosis which develop into female gametophytes. These give rise to egg cells. The gynoecium of a flower is also described using an alternative terminology wherein the structure one sees in the innermost whorl (consisting of an ovary, style and stigma) is called a pistil. A pistil may consist of a single carpel or a number of carpels fused together. The sticky tip of the pistil, the stigma, is the receptor of pollen. The supportive stalk, the style, becomes the pathway for pollen tubes to grow from pollen grains adhering to the stigma. The relationship to the gynoecium on the receptacle is described as hypogynous (beneath a superior ovary), perigynous (surrounding a superior ovary), or epigynous (above inferior ovary).

Structure

Although the arrangement described above is considered “typical”, plant species show a wide variation in floral structure. These modifications have significance in the evolution of flowering plants and are used extensively by botanists to establish relationships among plant species.

The four main parts of a flower are generally defined by their positions on the receptacle and not by their function. Many flowers lack some parts or parts may be modified into other functions and/or look like what is typically another part. In some families, like Ranunculaceae, the petals are greatly reduced and in many species the sepals are colorful and petal-like. Other flowers have modified stamens that are petal-like; the double flowers of Peonies and Roses are mostly petaloid stamens. Flowers show great variation and plant scientists describe this variation in a systematic way to identify and distinguish species.

Specific terminology is used to describe flowers and their parts. Many flower parts are fused together; fused parts originating from the same whorl are connate, while fused parts originating from different whorls are adnate; parts that are not fused are free. When petals are fused into a tube or ring that falls away as a single unit, they are sympetalous (also called gamopetalous). Connate petals may have distinctive regions: the cylindrical base is the tube, the expanding region is the throat and the flaring outer region is the limb. A sympetalous flower, with bilateral symmetry with an upper and lower lip, is bilabiate. Flowers with connate petals or sepals may have various shaped corolla or calyx, including campanulate, funnelform, tubular, urceolate, salverform or rotate.

Referring to "fusion," as it is commonly done, appears questionable because at least some of the processes involved may be non-fusion processes. For example, the addition of intercalary growth at or below the base of the primordia of floral appendages such as sepals, petals, stamens and carpels may lead to a common base that is not the result of fusion.

Left: A normal zygomorphic Streptocarpus flower. Right: An aberrant peloric Streptocarpus flower. Both of these flowers appeared on the Streptocarpus hybrid 'Anderson's Crows' Wings'.

Many flowers have a symmetry. When the perianth is bisected through the central axis from any point and symmetrical halves are produced, the flower is said to be actinomorphic or regular, e.g. rose or trillium. This is an example of radial symmetry. When flowers are bisected and produce only one line that produces symmetrical halves, the flower is said to be irregular or zygomorphic, e.g. snapdragon or most orchids.

Flowers may be directly attached to the plant at their base (sessile—the supporting stalk or stem is highly reduced or absent). The stem or stalk subtending a flower is called a peduncle. If a peduncle supports more than one flower, the stems connecting each flower to the main axis are called pedicels. The apex of a flowering stem forms a terminal swelling which is called the *torus* or receptacle.

Inflorescence

The familiar calla lily is not a single flower. It is actually an inflorescence of tiny flowers pressed together on a central stalk that is surrounded by a large petal-like bract.

In those species that have more than one flower on an axis, the collective cluster of flowers is termed an *inflorescence*. Some inflorescences are composed of many small flowers arranged in a formation that resembles a single flower. The common example of this is most members of the very large composite (Asteraceae) group. A single daisy or sunflower, for example, is not a flower but a flower *head*—an inflorescence composed of numerous flowers (or florets). An inflorescence may include specialized stems and modified leaves known as bracts.

Floral Diagrams and Floral Formulae

A *floral formula* is a way to represent the structure of a flower using specific letters, numbers and symbols, presenting substantial information about the flower in a compact form. It can represent a taxon, usually giving ranges of the numbers of different organs, or particular species. Floral formulae have been developed and their use has declined since. Prenner *et al.* (2010) devised an extension of the existing model to broaden the descriptive capability of the formula. The format of floral formulae differs in different parts of the world, yet they convey the same information.

The structure of a flower can also be expressed by the means of *floral diagrams*. The use of schematic diagrams can replace long descriptions or complicated drawings as a tool for understanding both floral structure and evolution. Such diagrams may show important features of flowers, including the relative positions of the various organs, including the presence of fusion and symmetry, as well as structural details.

Development

A flower develops on a modified shoot or *axis* from a determinate apical meristem (*determinate* meaning the axis grows to a set size). It has compressed internodes, bearing structures that in classical plant morphology are interpreted as highly modified leaves. Detailed developmental studies, however, have shown that stamens are often initiated more or less like modified stems (caulomes) that in some cases may even resemble branchlets. Taking into account the whole diversity in the development of the androecium of flowering plants, we find a continuum between modified leaves (phyllomes), modified stems (caulomes), and modified branchlets (shoots).

Flowering Transition

The transition to flowering is one of the major phase changes that a plant makes during its life cycle. The transition must take place at a time that is favorable for fertilization and the formation of seeds, hence ensuring maximal reproductive success. To meet these needs a plant is able to interpret important endogenous and environmental cues such as changes in levels of plant hormones and seasonable temperature and photoperiod changes. Many perennial and most biennial plants require vernalization to flower. The molecular interpretation of these signals is through the transmission of a complex signal known as florigen, which involves a variety of genes, including CONSTANS, FLOWERING LOCUS C and FLOWERING LOCUS T. Florigen is produced in the leaves in reproductively favorable conditions and acts in buds and growing tips to induce a number of different physiological and morphological changes.

The first step of the transition is the transformation of the vegetative stem primordia into floral primordia. This occurs as biochemical changes take place to change cellular differentiation of leaf, bud and stem tissues into tissue that will grow into the reproductive organs. Growth of the central part of the stem tip stops or flattens out and the sides develop protuberances in a whorled or spiral fashion around the outside of the stem end. These protuberances develop into the sepals, petals, stamens, and carpels. Once this process begins, in most plants, it cannot be reversed and the stems develop flowers, even if the initial start of the flower formation event was dependent of some environmental cue. Once the process begins, even if that cue is removed the stem will continue to develop a flower.

Yvonne Aitken has shown that flowering transition depends on a number of factors, and that plants flowering earliest under given conditions had the least dependence on climate whereas later-flowering varieties reacted strongly to the climate setup.

Organ Development

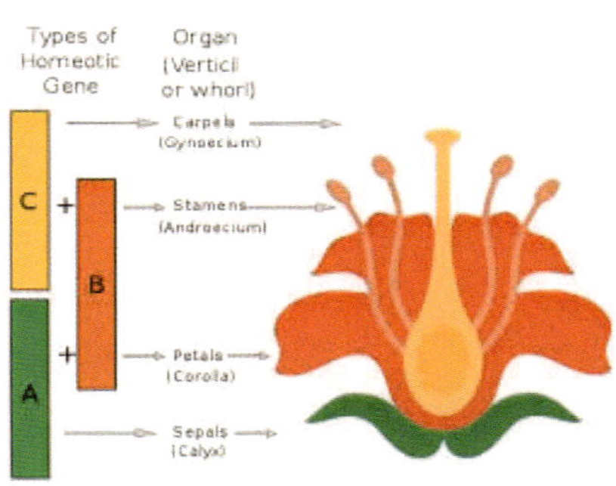

The ABC model of flower development

The molecular control of floral organ identity determination appears to be fairly well understood in some species. In a simple model, three gene activities interact in a combinatorial manner to determine the developmental identities of the organ primordia within the floral meristem. These gene functions are called A, B and C-gene functions. In the first floral whorl only A-genes are expressed, leading to the formation of sepals. In the second whorl both A- and B-genes are expressed, leading to the formation of petals. In the third whorl, B and C genes interact to form stamens and in the center of the flower C-genes alone give rise to carpels. The model is based upon studies of mutants in *Arabidopsis thaliana* and snapdragon, *Antirrhinum majus*. For example, when there is a loss of B-gene function, mutant flowers are produced with sepals in the first whorl as usual, but also in the second whorl instead of the normal petal formation. In the third whorl the lack of B function but presence of C-function mimics the fourth whorl, leading to the formation of carpels also in the third whorl.

Most genes central in this model belong to the MADS-box genes and are transcription factors that regulate the expression of the genes specific for each floral organ.

Floral Function

A "perfect flower", this *Crateva religiosa* flower has both stamens (outer ring) and a pistil (center).

The principal purpose of a flower is the reproduction of the individual and the species. All flowering plants are *heterosporous*, producing two types of spores. Microspores are produced by meiosis inside anthers while megaspores are produced inside ovules, inside an ovary. In fact, anthers typically consist of four microsporangia and an ovule is an integumented megasporangium. Both types of spores develop into gametophytes inside sporangia. As with all heterosporous plants, the gametophytes also develop inside the spores (are endosporic).

In the majority of species, individual flowers have both functional carpels and stamens. Botanists describe these flowers as being *perfect* or *bisexual* and the species as *hermaphroditic*. Some flowers lack one or the other reproductive organ and called *imperfect* or *unisexual*. If unisex flowers are found on the same individual plant but in different locations, the species is said to be *monoecious*. If each type of unisex flower is found only on separate individuals, the plant is *dioecious*.

Flower Specialization and Pollination

Flowering plants usually face selective pressure to optimize the transfer of their pollen, and this is typically reflected in the morphology of the flowers and the behaviour of the plants. Pollen may be transferred between plants via a number of 'vectors'. Some plants make use of abiotic vectors — namely wind (anemophily) or, much less commonly, water (hydrophily). Others use biotic vectors including insects (entomophily), birds (ornithophily), bats (chiropterophily) or other animals. Some plants make use of multiple vectors, but many are highly specialised.

Cleistogamous flowers are self-pollinated, after which they may or may not open. Many Viola and some Salvia species are known to have these types of flowers.

The flowers of plants that make use of biotic pollen vectors commonly have glands called nectaries that act as an incentive for animals to visit the flower. Some flowers have patterns, called nectar guides, that show pollinators where to look for nectar. Flowers also attract pollinators by scent and color. Still other flowers use mimicry to attract pollinators. Some species of orchids, for example, produce flowers resembling female bees in color, shape, and scent. Flowers are also specialized in shape and have an arrangement of the stamens that ensures that pollen grains are transferred to the bodies of the pollinator when it lands in search of its attractant (such as nectar, pollen, or a mate). In pursuing this attractant from many flowers of the same species, the pollinator transfers pollen to the stigmas—arranged with equally pointed precision—of all of the flowers it visits.

Anemophilous flowers use the wind to move pollen from one flower to the next. Examples include grasses, birch trees, ragweed and maples. They have no need to attract pollinators and therefore tend not to be "showy" flowers. Male and female reproductive organs are generally found in separate flowers, the male flowers having a number of long filaments terminating in exposed stamens, and the female flowers having long, feather-like stigmas. Whereas the pollen of animal-pollinated flowers tends to be large-grained, sticky, and rich in protein (another "reward" for pollinators), anemophilous flower pollen is usually small-grained, very light, and of little nutritional value to animals.

Pollination

Grains of pollen sticking to this bee will be transferred to the next flower it visits

The primary purpose of a flower is reproduction. Since the flowers are the reproductive organs of plant, they mediate the joining of the sperm, contained within pollen, to the ovules — contained in the ovary. Pollination is the movement of pollen from the anthers to the stigma. The joining of the sperm to the ovules is called fertilization. Normally pollen is moved from one plant to another, but many plants are able to self pollinate. The fertilized ovules produce seeds that are the next generation. Sexual reproduction produces genetically unique offspring, allowing for adaptation. Flowers have specific designs which encourages the transfer of pollen from one plant to another of the same species. Many plants are dependent upon external factors for pollination, including: wind and animals, and especially insects. Even large animals such as birds, bats, and pygmy possums can be employed. The period of time during which this process can take place (the flower is fully expanded and functional) is called *anthesis*. The study of pollination by insects is called *anthecology*.

Pollination Mechanism

The pollination mechanism employed by a plant depends on what method of pollination is utilized.

Most flowers can be divided between two broad groups of pollination methods:

Entomophilous: flowers attract and use insects, bats, birds or other animals to transfer pollen from one flower to the next. Often they are specialized in shape and have an arrangement of the stamens that ensures that pollen grains are transferred to the bodies of the pollinator when it lands in search of its attractant (such as nectar, pollen, or a mate). In pursuing this attractant from many flowers of the same species, the pollinator transfers pollen to the stigmas—arranged with equally pointed precision—of all of the flowers it visits. Many flowers rely on simple proximity between flower parts to ensure pollination. Others, such as the *Sarracenia* or lady-slipper orchids, have elaborate designs to ensure pollination while preventing self-pollination.

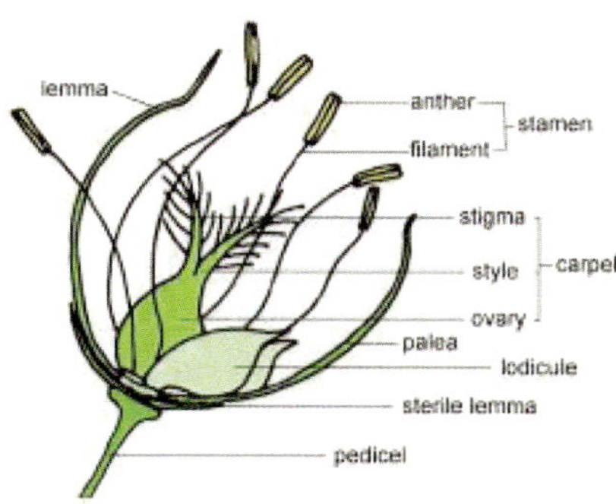

Grass flower with vestigial perianth or lodicules

Anemophilous: flowers use the wind to move pollen from one flower to the next, examples include the grasses, Birch trees, Ragweed and Maples. They have no need to attract pollinators and therefore tend not to grow large blossoms. Whereas the pollen of entomophilous flowers tends to be large-grained, sticky, and rich in protein (another "reward" for pollinators), anemophilous flower pollen is usually small-grained, very light, and of little nutritional value to insects, though it may still be gathered in times of dearth. Honeybees and bumblebees actively gather anemophilous corn (maize) pollen, though it is of little value to them.

Some flowers with both stamens and a pistil are capable of self-fertilization, which does increase the chance of producing seeds but limits genetic variation. The extreme case of self-fertilization occurs in flowers that always self-fertilize, such as many dandelions. Some flowers are self-pollinated and use flowers that never open or are self-pollinated before the flowers open, these flowers are called cleistogamous. Many Viola species and some Salvia have these types of flowers. Conversely, many species of plants have ways of preventing self-fertilization. Unisexual male and female flowers on the same plant may not appear or mature at the same time, or pollen from the same plant may be incapable of fertilizing its ovules. The latter flower types, which have chemical barriers to their own pollen, are referred to as self-sterile or self-incompatible.

Attraction Methods

A Bee orchid has evolved over many generations to better mimic a female bee to attract male bees as pollinators.

Plants cannot move from one location to another, thus many flowers have evolved to attract animals to transfer pollen between individuals in dispersed populations. Flowers that are insect-pollinated are called *entomophilous*; literally "insect-loving" in Greek. They can be highly modified along with the pollinating insects by co-evolution. Flowers commonly have glands called *nectaries* on various parts that attract animals looking for nutritious nectar. Birds and bees have color vision, enabling them to seek out "colorful" flowers.

Some flowers have patterns, called nectar guides, that show pollinators where to look for nectar; they may be visible only under ultraviolet light, which is visible to bees and some other insects. Flowers also attract pollinators by scent and some of those scents are pleasant to our sense of smell. Not all flower scents are appealing to humans; a number of flowers are pollinated by insects that are attracted to rotten flesh and have flowers that smell like dead animals, often called Carrion flowers, including *Rafflesia*, the titan arum, and the North American pawpaw (*Asimina triloba*). Flowers pollinated by night visitors, including bats and moths, are likely to concentrate on scent to attract pollinators and most such flowers are white.

Other flowers use mimicry to attract pollinators. Some species of orchids, for example, produce flowers resembling female bees in color, shape, and scent. Male bees move from one such flower to another in search of a mate.

Flower-pollinator Relationships

Many flowers have close relationships with one or a few specific pollinating organisms. Many flowers, for example, attract only one specific species of insect, and therefore rely on that insect for successful reproduction. This close relationship is often given as an example of coevolution, as the flower and pollinator are thought to have developed together over a long period of time to match each other's needs.

This close relationship compounds the negative effects of extinction. The extinction of either member in such a relationship would mean almost certain extinction of the other member as well. Some endangered plant species are so because of shrinking pollinator populations.

Pollen Allergy

There is much confusion about the role of flowers in allergies. For example, the showy and entomophilous goldenrod (*Solidago*) is frequently blamed for respiratory allergies, of which it is innocent, since its pollen cannot be airborne. The types of pollen that most commonly cause allergic reactions are produced by the plain-looking plants (trees, grasses, and weeds) that do not have showy flowers. These plants make small, light, dry pollen grains that are custom-made for wind transport.

The type of allergens in the pollen is the main factor that determines whether the pollen is likely to cause hay fever. For example, pine tree pollen is produced in large amounts by a common tree, which would make it a good candidate for causing allergy. It is, however, a relatively rare cause of allergy because the types of allergens in pine pollen appear to make it less allergenic. Instead the allergen is usually the pollen of the contemporary bloom of anemophilous ragweed (*Ambrosia*), which can drift for many miles. Scientists have collected samples of ragweed pollen 400 miles out at sea and 2 miles high in the air. A single ragweed plant can generate a million grains of pollen per day.

Among North American plants, weeds are the most prolific producers of allergenic pollen. Ragweed is the major culprit, but other important sources are sagebrush, redroot pigweed, lamb's quarters, Russian thistle (tumbleweed), and English plantain.

It is common to hear people say they are allergic to colorful or scented flowers like roses. In fact, only florists, gardeners, and others who have prolonged, close contact with flowers are likely to be sensitive to pollen from these plants. Most people have little contact with the large, heavy, waxy pollen grains of such flowering plants because this type of pollen is not carried by wind but by insects such as butterflies and bees.

Seed Dispersal

Evolution

Archaefructus liaoningensis, one of the earliest known flowering plants

While land plants have existed for about 425 million years, the first ones reproduced by a simple adaptation of their aquatic counterparts: spores. In the sea, plants—and some animals—can simply scatter out genetic clones of themselves to float away and grow elsewhere. This is how early plants reproduced. But plants soon evolved methods of protecting these copies to deal with drying out and other damage which is even more likely on land than in the sea. The protection became the seed, though it had not yet evolved the flower. Early seed-bearing plants include the ginkgo and conifers.

Several groups of extinct gymnosperms, particularly seed ferns, have been proposed as the ancestors of flowering plants but there is no continuous fossil evidence showing exactly how flowers evolved. The apparently sudden appearance of relatively modern flowers in the fossil record posed such a problem for the theory of evolution that it was called an "abominable mystery" by Charles Darwin. Recently discovered angiosperm fossils such as *Archaefructus*, along with further discoveries of fossil gymnosperms, suggest how angiosperm characteristics may have been acquired in a series of steps. An early fossil of a flowering plant, *Archaefructus liaoningensis* from China, is dated about 125 million years old. Even earlier from China is the 125–130 million years old *Archaefructus sinensis*. Now, another plant (130 million-year-old *Montsechia vidalii*, discovered in Spain) takes the title of world's oldest flower from *Archaefructus sinensis*.

Recent DNA analysis (molecular systematics) shows that *Amborella trichopoda*, found on the Pacific island of New Caledonia, is the only species in the sister group to the rest of the flowering plants, and morphological studies suggest that it has features which may have been characteristic of the earliest flowering plants.

Amborella trichopoda may have characteristic features of the earliest flowering plants

While there is only hard proof of such flowers existing about 130 million years ago, there is some circumstantial evidence that they did exist up to 250 million years ago. A chemical used by plants to defend their flowers, oleanane, has been detected in fossil plants that old, including gigantopterids, which evolved at that time and bear many of the traits of modern, flowering plants, though they are not known to be flowering plants themselves, because only their stems and prickles have been found preserved in detail; one of the earliest examples of petrification.

In August 2017, scientists presented a detailed description and 3D model image of possibly the first flower that lived about 140 million years ago.

The similarity in leaf and stem structure can be very important, because flowers are genetically just an adaptation of normal leaf and stem components on plants, a combination of genes normally responsible for forming new shoots. The most primitive flowers are thought to have had a variable number of flower parts, often separate from (but in contact with) each other. The flowers would have tended to grow in a spiral pattern, to be bisexual (in plants, this means both male and female parts on the same flower), and to be dominated by the ovary (female part). As flowers grew more advanced, some variations developed parts fused together, with a much more specific number and design, and with either specific sexes per flower or plant, or at least "ovary inferior".

The general assumption is that the function of flowers, from the start, was to involve animals in the reproduction process. Pollen can be scattered without bright colors and obvious shapes, which would therefore be a liability, using the plant's resources, unless they provide some other benefit. One proposed reason for the sudden, fully developed appearance of flowers is that they evolved in an isolated setting like an island, or chain of islands, where the plants bearing them were able to develop a highly specialized relationship with some specific animal (a wasp, for example), the way many island species develop today. This symbiotic relationship, with a hypothetical wasp bearing pollen from one plant to another much the way fig wasps do today, could have eventually resulted in both the plant(s) and their partners developing a high degree of specialization. Island genetics is believed to be a common source of speciation, especially when it comes to radical adaptations which seem to have required inferior transitional forms. Note that the wasp example is not incidental; bees, apparently evolved specifically for symbiotic plant relationships, are descended from wasps.

Likewise, most fruit used in plant reproduction comes from the enlargement of parts of the flower. This fruit is frequently a tool which depends upon animals wishing to eat it, and thus scattering the seeds it contains.

While many such symbiotic relationships remain too fragile to survive competition with mainland organisms, flowers proved to be an unusually effective means of production, spreading (whatever their actual origin) to become the dominant form of land plant life.

Flower evolution continues to the present day; modern flowers have been so profoundly influenced by humans that many of them cannot be pollinated in nature. Many modern, domesticated flowers used to be simple weeds, which only sprouted when the ground was disturbed. Some of them tended to grow with human crops, and the prettiest did not get plucked because of their beauty, developing a dependence upon and special adaptation to human affection.

Color

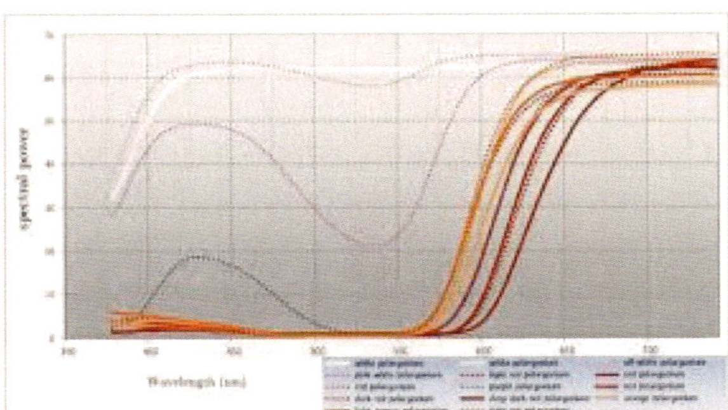

Reflectance spectra for the flowers of several varieties of rose. A red rose absorbs about 99.7% of light across a broad area below the red wavelengths of the spectrum, leading to an exceptionally *pure* red. A yellow rose will reflect about 5% of blue light, producing an unsaturated yellow (a yellow with a degree of white in it).

Many flowering plants reflect as much light as possible within the range of visible wavelengths of the pollinator the plant intends to attract. Flowers that reflect the full range of visible light are generally perceived as *white* by a human observer. An important feature of white flowers is that they reflect equally across the visible spectrum. While many flowering plants use white to attract pollinators, the use of color is also widespread (even within the same species). Color allows a flowering plant to be more specific about the pollinator it seeks to attract. The color model used by human color reproduction technology (CMYK) relies on the modulation of pigments that divide the spectrum into broad areas of absorption. Flowering plants by contrast are able to shift the transition point wavelength between absorption and reflection. If it is assumed that the visual systems of most pollinators view the visible spectrum as circular then it may be said that flowering plants produce color by absorbing the light in one region of the spectrum and reflecting the light in the other region. With CMYK, color is produced as a function of the amplitude of the broad regions of absorption. Flowering plants by contrast produce color by modifying the frequency (or rather wavelength) of the light reflected. Most flowers absorb light in the blue to yellow region of the spectrum and reflect light from the green to red region of the spectrum. For many species of flowering plant, it is the transition point that characterizes the color that they produce. Color may be modulated by shifting the transition point between absorption and reflection and in this way a flowering plant may specify which pollinator it seeks to attract. Some flowering plants also have a limited ability to modulate areas of absorption. This is typically not as precise as control over wavelength. Humans observers will perceive this as degrees of saturation (the amount of *white* in the color).

Symbolism

Lilies are often used to denote life or resurrection

Many flowers have important symbolic meanings in Western culture. The practice of assigning meanings to flowers is known as floriography. Some of the more common examples include:

- Red roses are given as a symbol of love, beauty, and passion.
- Poppies are a symbol of consolation in time of death. In the United Kingdom, New Zealand, Australia and Canada, red poppies are worn to commemorate soldiers who have died in times of war.
- Irises/Lily are used in burials as a symbol referring to "resurrection/life". It is also associated with stars (sun) and its petals blooming/shining.
- Daisies are a symbol of innocence.

Flowers are common subjects of still life paintings, such as this one by Ambrosius Bosschaert the Elder

Because of their varied and colorful appearance, flowers have long been a favorite subject of visual artists as well. Some of the most celebrated paintings from well-known painters are of flowers, such as Van Gogh's sunflowers series or Monet's water lilies. Flowers are also dried, freeze dried and pressed in order to create permanent, three-dimensional pieces of flower art.

Flowers within art are also representative of the female genitalia, as seen in the works of artists such as Georgia O'Keeffe, Imogen Cunningham, Veronica Ruiz de Velasco, and Judy Chicago, and in fact in Asian and western classical art. Many cultures around the world have a marked tendency to associate flowers with femininity.

The great variety of delicate and beautiful flowers has inspired the works of numerous poets, especially from the Romantic era. Famous examples include William Wordsworth's *I Wandered Lonely as a Cloud* and William Blake's *Ah! Sun-Flower*.

Their symbolism in dreams has also been discussed, with possible interpretations including "blossoming potential".

The Roman goddess of flowers, gardens, and the season of Spring is Flora. The Greek goddess of spring, flowers and nature is Chloris.

In Hindu mythology, flowers have a significant status. Vishnu, one of the three major gods in the Hindu system, is often depicted standing straight on a lotus flower. Apart from the association with Vishnu, the Hindu tradition also considers the lotus to have spiritual significance. For example, it figures in the Hindu stories of creation.

Floral Biology

Floral biology is an area of ecological research that studies the evolutionary factors that have moulded the structures, behaviour and physiological aspects involved in the flowering of plants. The field is broad and interdisciplinary and involves research requiring expertise from multiple disciplines that can include botany, ethology, biochemistry, entomology and a range of analytical techniques. A slightly narrower area of research within floral biology is sometimes termed as pollination biology or anthecology.

Flowers are structures that are produced by angiosperms and their evolution is intricately associated with their pollinators, particularly insects. Flowers are costly structures that target pollinators by offering them rewards so as to enhance cross-pollination. The evolution of the size of flowers, their structure and the nature of rewards and the way these signals are transmitted and perceived by potential pollinators are typically examined in terms of the costs incurred and the benefits accrued. The timing of flowering, the duration of flowering and the mode by which they cease to function once their role is fulfilled are all areas of research within the field of floral biology.

Studies in floral biology can have applications since pollination and fruit set are key factors that affect yield in all crop plants.

Aspects

Flowering plants, angiosperms, are relatively recent among the plants. The oldest flower structures date to about 140 million years ago. Flowering plants underwent a major diversification after this period. Darwin saw this as an "abominable mystery" in a letter to Joseph Hooker in 1879. The earliest flowers were principally actinomorphic or having radial symmetry with multiple axes of symmetry. From these evolved flowers with bilateral symmetry or zygomorphy. It is thought that the attraction of insect pollinators led by visual cues had an influence in the evolution of zygomorphy.

The earliest groups of flowering plants among the Magnoliids and the families Choranthaceae, Ceratophyllaceae, Nymphaeaceae, Annonaceae, and Aristolochiaceae are bisexual with both male and female parts present and functional within the usually large floral structure. The main pollinators of these flowers were beetles, flies and thrips. They evolved mechanisms to reduce self-pollination by changing the timing of maturity of the male and female parts. This altered timing mechanism or dichogamy was principally expressed by protogyny or the early maturation of the female parts and only rarely by protandry or the early maturation of the male parts. The early flowers were principally flat and dish-like with the evolution of deeper corolla tubes being a later innovation and principally associated with long-tongued pollinators such as moths. There are a large number of other mechanisms that enhance cross-pollination and prevent self-pollination. The forces that lead to the evolution of such systems as bearing male and female flowers on separate kinds of plants is still unclear.

The evolution from solitary flowers to the production of inflorescences is also thought to be influenced by pollinator behaviour. Clusters of flowers may increase the visitation rates of pollinators. It may also reduce the risk of damage to individual flowers.

The production of volatile chemicals by flowers is targeted towards insects. Some evidence shows that there is significant overlap between the chemicals produced by plants and those used by insects for their communications, especially for mating. In the classic case of orchids in the genus *Ophrys*, the volatiles mimic the female sex pheromone of bees which attempt to copulate with the flower and thereby pollinate them. A study of the evolution of volatile chemicals in scarab beetles and flowers that attract them in the family Araceae showed that the insects had evolved the chemicals in the Jurassic Period while the plants evolved the attractants later in the Cretaceous Period.

The colours of flowers are another area of enquiry. Some distinct patterns have been noted, for instance bird-pollinated flowers are predominantly red while night-flowering plants tend to be white. The colours of flowers are produced by a variety of pigmentary mechanisms and they are meant to signal messages to pollinators. Flowers that have been pollinated often quickly wither and the nutrients resorbed by the plant. In some case parts of the flower may undergo colour changes to indicate their being unfit for visitation by pollinators.

Floral Formula

Floral formula is a means to represent the structure of a flower using numbers, letters and various symbols, presenting substantial information about the flower in a compact form. It can represent particular species, or can be generalized to characterize higher taxa, usually giving ranges of organ numbers. Floral formulae are one of the two ways of describing flower structure developed, the other being floral diagrams. The format of floral formulae differs between authors, yet they tend to convey the same information.[1

Contained Information

Organ Numbers and Fusion

The formula expresses counts of different floral organs, these are usually preceded by letters or abbreviations according to the organ type. They are ordered corresponding to the arrangement of the parts of the flower from the outside to the inside:

<table>
<tr><th>Bracts</th><th>Bracteoles</th><th colspan="2">Tepals (perigon or perianth), or sepals (calyx) and petals (corolla)</th><th>Stamens (androecium)</th><th>Carpels (gynoecium)</th><th>Ovules</th></tr>
<tr><td rowspan="2">B</td><td rowspan="2">Bt</td><td colspan="2">P or CaCo</td><td rowspan="2">A</td><td rowspan="2">G</td><td rowspan="2">V or O</td></tr>
<tr><td>K or Ca</td><td>C or Co</td></tr>
</table>

The labels with darker backgrounds are less common. "V" used by Prenner *et al.* for the number of ovules per gynoecium is followed by lowercase letter describing the type of placentation. For epicalyx/calyculus, the letter "k" is used.

The numbers are inserted after the labels, they may be formatted as sub- or superscript. If an organ is absent, its number is written as "0" or it is omitted, if there are "many" (usually more than 10–12) instances, it can be written as "∞". Whorls of the same organ are separated by "+". Organ counts within a whorl can be separated by ":", for example when part of the whorl is morphologically different. A range can be given if the number is variable, e.g. when the formula summarizes a taxon.

> K_{3+3} – a calyx with six free sepals, arranged as two separate whorls.
>
> $A\infty$ – many stamens.
>
> P3–12 – perianth from three to twelve tepals.

Groups of organs can be described by writing the number of instances in the group as superscript.

> $A5^2$ – 5 groups of 2 stamens.

The formula can also express organ fusion. Fusion of one organ type can be shown by enclosing the number in a circle, fusion of different organs can be represented by ties, as e.g. in Judd *et al.* Prenner *et al.* state that this method is difficult to achieve via standard typesetting. Joining of organs can be more readily written using parentheses "(...)" if instances of the same organ are fused. Fusion between different organs can be achieved by square "[...]", eventually curly brackets "{...}".

> A(5) – five fused stamens.
>
> [C(5) A5] – corolla fused from 5 petals, fused to stamens.

Prenner *et al.* propose superscript zero for a lost organ, and superscript "r" for a reduced one. Ronse De Craene uses a degree symbol to mark a staminode (infertile stamen) or pistillode (infertile carpel).

> $A3{:}2^r{+}5^o$ – (Prenner *et al.*) androecium in two whorls, first contains 3 stamens and 2 staminodes, second whorl lost.
>
> $A1{+}2^o$ – (Ronse De Craene) androecium in two whorls, first whorl containing a stamen, second whorl containing two staminodes.

Ovary Position

Ovary position is shown by alternating the "G" label. Simpson circumvents the intricate formatting by expressing the ovary position by words.

	superior ovary	**inferior ovary**	**half-inferior ovary**
Prenner *et al.*, Ronse De Craene	$\underline{G}$	Ĝ, Ğ	-G-
Sattler		G	~~G~~
Simpson	G..., superior	G..., inferior	G..., half-inferior

Symmetry

Symmetry or arrangement may be described for the whole flower, in such case the corresponding symbol is usually placed at the beginning of the formula. It may be also outlined separately for different organs, placing it after their labels or numbers, or it may not be included in the formula at all. It is described by following symbols:

	polysymmetry (actinomorphic)	disymmetry	monosymmetry (zygomorphic)	asymmetry	spiral arrangement
Prenner *et al.*	*	+	↓, → or Ø, depending on the symmetry plane orientation	∂	not mentioned
Ronse De Craene	✶	↔	↓, arrow orientation depending on symmetry plane orientation	↯	↺
Sattler	✳	+	·|·		
Judd *et al.*	*		X	$	not mentioned
Subrahmanyam	⊕	not mentioned	% in median plane, ÷ in lateral plane	not mentioned	
Rosypal	✳	⤧	↓	↯	๑

Sexuality

Sexuality of the flower can be highlighted by ☿ or ⚥ for hermaphrodite (bisexual), ♂ for male (staminate) and ♀ for female (pistillate) flower. The symbols are usually placed at the beginning of the formula, after or before the symmetry symbol. Prenner *et al.* recommend to use the corresponding symbols (♀ and ♂) only for flowers of separate sexuality. Ronse De Craene utilizes the words "pistillate" or "staminate" instead of the symbols.

Floral formula can also incorporate the fruit type, Judd *et al.* place it at the very end.

Examples

↯ K3 [C3 A1°–3°+½:2°] Ğ(3) – the formula of *Canna edulis*; asymmetric flower; calyx of three free sepals; corolla of three free petals joined with androecium; androecium in two whorls, the outer whorl contains 1–3 staminodes, the inner contains ½ of a stamen and 2 staminodes; gynoecium fused of 3 carpels, inferior ovary.

B Bt^{c} $K3{:}(2)^{c}$↓ $C3{:}2^{r}$↓ $A(3){:}2^{r}$↓$+4^{r}{:}1^{o}$ G1↓ Vm8–10 – the formula of *Tamarindus indica*; bract and petaloid bracteoles; monosymmetric calyx of three and two petaloid sepals; monosymmetric corolla of three and two reduced petals; two whorls of stamens, the outer monosymmetric from three fused and two reduced stamens, the inner of 4 reduced and 1 lost stamen; monosymmetric gynoecium of 1 carpel with superior ovary; marginal placentation with 8–10 ovules per gynoecium.

ABC Model of Flower Development

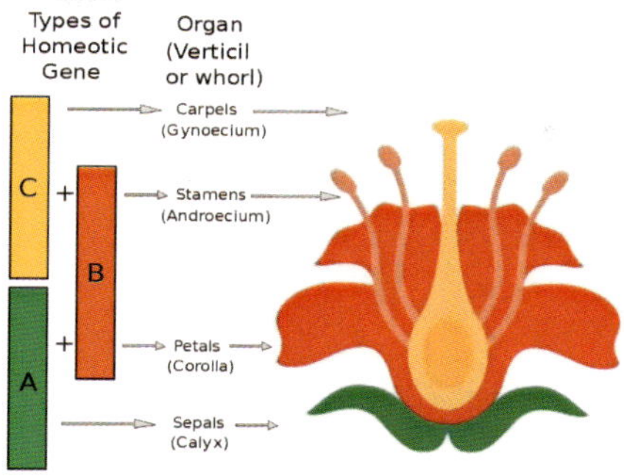

ABC model of flower development guided by three groups of homeotic genes.

The ABC model of flower development is a scientific model of the process by which flowering plants produce a pattern of gene expression in meristems that leads to the appearance of an organ oriented towards sexual reproduction, a flower. There are three physiological developments that must occur in order for this to take place: firstly, the plant must pass from sexual immaturity into a sexually mature state (i.e. a transition towards flowering); secondly, the transformation of the apical meristem's function from a vegetative meristem into a floral meristem or inflorescence; and finally the growth of the flower's individual organs. The latter phase has been modelled using the ABC model, which aims to describe the biological basis of the process from the perspective of molecular and developmental genetics.

An external stimulus is required in order to trigger the differentiation of the meristem into a flower meristem. This stimulus will activate mitotic cell division in the meristem, particularly on its sides where new primordia are formed. This same stimulus will also cause the meristem to follow a developmental pattern that will lead to the growth of floral meristems as opposed to vegetative meristems. The main difference between these two types of meristem, apart from the obvious disparity between the objective organ, is the verticillate (or whorled) phyllotaxis, that is, the absence of stem elongation among the successive whorls or verticils of the primordium. These verticils follow an acropetal development, giving rise to sepals, petals, stamens and carpels. Another difference from vegetative axillary meristems is that the floral meristem is «determined», which means that, once differentiated, its cells will no longer divide.

The identity of the organs present in the four floral verticils is a consequence of the interaction of at least three types of gene products, each with distinct functions. According to the ABC model, functions A and C are required in order to determine the identity of the verticils of the perianth and the reproductive verticils, respectively. These functions are exclusive and the absence of one of them means that the other will determine the identity of all the floral verticils. The B function allows the differentiation of petals from sepals in the secondary verticil, as well as the differentiation of the stamen from the carpel on the tertiary verticil.

Goethe's foliar theory was formulated and it suggests that the constituent parts of a flower are structurally modified leaves, which are functionally specialized for reproduction or protection. The theory was first published in 1790 in the essay "Metamorphosis of Plants" ("*Versuch die Metamorphose der Pflanzen zu erklären*"). where Goethe wrote:

"...we may equally well say that a stamen is a contracted petal, as that a petal is a stamen in a state of expansion; or that a sepal is a contracted stem leaf approaching a certain stage of refinement, as that a stem leaf is a sepal expanded by the influx of cruder saps".

Floral Transition

The transition from the vegetative phase to a reproductive phase involves a dramatic change in the plant's vital cycle, perhaps the most important one, as the process must be carried out correctly in order to guarantee that the plant produces descendents. This transition is characterised by the induction and development of the meristem of the inflorescence, which will produce a collection of flowers or one flower, where only one is produced. This morphogenetic change contains both endogenous and exogenous elements: For example, in order for the change to be initiated the plant must have a certain number of leaves and contain a certain level of total biomass. Certain environmental conditions are also required such as a characteristic photoperiod. Plant hormones play an important part in the process, with the gibberellins having a particularly important role.

There are many signals that regulate the molecular biology of the process. The following three genes in *Arabidopsis thaliana* possess both common and independent functions in floral transition: *FLOWERING LOCUS T* (*FT*), *LEAFY* (*LFY*), *SUPPRESSOR OF OVEREXPRESSION OF CONSTANS1* (*SOC1*, also called *AGAMOUS-LIKE20*). *SOC1* is a MADS-box-type gene, which integrates responses to photoperiod, vernalization and gibberellins.

Formation of the Floral Meristem or the Inflorescence

The meristem can be defined as the tissue or group of plant tissues that contain undifferentiated stem cells, which are capable of producing any type of cell tissue. Their maintenance and development, both in the vegetative meristem or the meristem of the inflorescence is controlled by genetic cell fate determination mechanisms. This means that a number of genes will directly regulate, for example, the maintenance of the stem cell's characteristics (gene *WUSCHEL* or *WUS*), and others will act via negative feedback mechanisms in order to inhibit a characteristic (gene *CLAVATA* or *CLV*). In this way both mechanisms give rise to a feedback loop, which along with other elements lend a great deal of robustness to the system. Along with the *WUS* gene the *SHOOT-MERISTEMLESS* (*STM*) gene also represses the differentiation of the meristematic dome. This gene acts by inhibiting the possible differentiation of the stem cells but still allows cell division in the daughter cells, which, had they been allowed to differentiate, would have given rise to distinct organs.

Floral Architecture

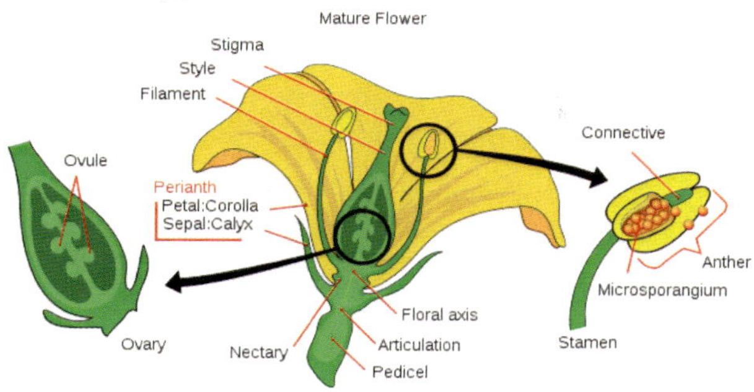

Anatomy of a flower

A flower's anatomy, as defined by the presence of a series of organs (sepals, petals, stamens and carpels) positioned according to a given pattern, facilitate sexual reproduction in flowering plants. The flower arises from the activity of three classes of genes, which regulate floral development: genes which regulate the identity of the meristem, the identity of the flower organ and finally cadastral genes.

- Meristem identity genes. Code for the transcription factors required to initiate the induction of the identity genes. They are positive regulators of organ identity during floral development.
- Organ identity genes. Directly control organ identity and also code for transcription factors that control the expression of other genes, whose products are implicated in the formation or function of the distinct organs of the flower.
- Cadastral genes. Act as spatial regulators for the organ identity genes by defining boundaries for their expression. In this way they control the extent to which genes interact thereby regulating whether they act in the same place at the same time.

The ABC Model

The ABC model of flower development was first formulated by George Haughn and Chris Somerville in 1988. It was first used as a model to describe the collection of genetic mechanisms that establish floral organ identity in the Rosids, as exemplified by *Arabidopsis thaliana*, and the Asterids, as demonstrated by *Antirrhinum majus*. Both species have four verticils (sepals, petals, stamens and carpels), which are defined by the differential expression of a number of homeotic genes present in each verticil. This means that the sepals are solely characterized by the expression of A genes, while the petals are characterized by the co-expression of A and B genes. The B and C genes establish the identity of the stamens and the carpels only require C genes to be active. It should be noted that type A and C genes are reciprocally antagonistic.

The fact that these homeotic genes determine an organ's identity becomes evident when a gene that represents a particular function, for example the A gene, is not expressed. In *Arabidopsis* this loss results in a flower which is composed of one verticil of carpels, another containing stamens and another of carpels. This method for studying gene function uses reverse genetics techniques to produce transgenic plants that contain a mechanism for gene silencing through RNA interference. In other studies, using forward genetics techniques such as genetic mapping, it is the analysis of the phenotypes of flowers with structural anomalies that leads to the cloning of the gene of interest. The flowers may possess a non-functional or over expressed allele for the gene being studied.

The existence of two supplementary functions, D and E, have also been proposed in addition to the A, B and C functions already discussed. Function D specifies the identity of the ovule, as a separate reproductive function from the development of the carpels, which occurs after their determination. Function E relates to a physiological requirement that is a characteristic of all floral verticils, although, it was initially described as necessary for the development of the three innermost verticils (Function E *sensu stricto*). However, its broader definition (*sensu lato*) suggests that it is required in the four verticils. Therefore, when Function D is lost the structure of the ovules becomes similar to that of leaves and when Function E is lost *sensu stricto*, the floral organs of the three outer most verticils are transformed into sepals, while on losing Function E *sensu lato*, all the verticils are similar to leaves. It is interesting to note that the gene products of genes with D and E functions are also MADS-box genes.

Genetic Analysis

Flower of *A. thaliana.*

Flowers of *A. majus.*

Flowers of *Petunia hybrid.*

The methodology for studying flower development involves two steps. Firstly, the identification of the exact genes required for determining the identity of the floral meristem. In *A. thaliana* these include APETALA1 (*AP1*) and LEAFY (*LFY*). Secondly, genetic analysis is carried out on the aberrant phenotypes for the relative characteristics of the flowers, which allows the characterization of the homeotic genes implicated in the process.

Analysis of Mutants

There are a great many mutations that affect floral morphology, although the analysis of these mutants is a recent development. Supporting evidence for the existence of these mutations comes from the fact that a large number affect the identity of floral organs. For example, some organs develop in a location where others should develop. This is called homeotic mutation, which is analogous to HOX gene mutations found in *Drosophila*. In *Arabidopsis* and *Antirrhinum*, the two taxa on which models are based, these mutations always affect adjacent verticils. This allows the characterization of three classes of mutation, according to which verticils are affected:

- Mutations in type A genes, these mutations affect the calyx and corolla, which are the outermost verticils. In these mutants, such as APETALA2 in *A. thaliana*, carpels develop instead of sepals and stamen in place of petals. This means that, the verticils of the perianth are transformed into reproductive verticils.
- Mutations in type B genes, these mutations affect the corolla and the stamen, which are the intermediate verticils. Two mutations have been found in *A. thaliana*, APETALA3 and PISTILLATA, which cause development of sepals instead of petals and carpels in the place of stamen.
- Mutations in type C genes, these mutations affect the reproductive verticils, namely the stamen and the carpels. The *A. thaliana* mutant of this type is called AGAMOUS, it possesses a phenotype containing petals instead of stamen and sepals instead of carpels.

Techniques for Detecting Differential Expression

Cloning studies have been carried out on DNA in the genes associated with the affected homeotic functions in the mutants discussed above. These studies used serial analysis of gene expression throughout floral development to show patterns of tissue expression, which, in general, correspond with the predictions of the ABC model.

The nature of these genes corresponds to that of transcription factors, which, as expected, have analogous structures to a group of factors contained in yeasts and animal cells. This group is called MADS, which is an acronym for the different factors contained in the group. These MADS factors have been detected in all the vegetable species studied, although the involvement of other elements involved in the regulation of gene expression cannot be discounted.

Genes Exhibiting type-A Function

In *A. thaliana*, function A is mainly represented by two genes *APETALA1* (*AP1*) and *APETALA2* (*AP2*) *AP1* is a MADS-box type gene, while *AP2* belongs to the family of genes that contains AP2, which it gives its name to and which consists of transcription factors that are only found in plants. AP2 has also been shown to complex with the co-repressor TOPLESS (TPL) in developing floral buds to repress the C-class gene *AGAMOUS (AG)*. However, *AP2* is not expressed in the shoot apical meristem (SAM), which contains the latent stem cell population throughout the adult life of *Arabidopsis*, and so it is speculated that TPL works with some other A-class gene in the SAM to repress *AG.AP1* functions as a type A gene, both in controlling the identity of sepals and petals, and it also acts in the floral meristem. *AP2* not only functions in the first two verticils, but also in the remaining two, in developing ovules and even in leaves. It is also likely that post-transcriptional regulation exists, which controls it's A function, or even that it has other purposes in the determination of organ identity independent of that mentioned here.

In *Antirrhinum*, the orthologous gene to *AP1* is *SQUAMOSA* (*SQUA*), which also has a particular impact on the floral meristem. The homologs for *AP2* are *LIPLESS1* (*LIP1*) and *LIPLESS2* (*LIP2*), which have a redundant function and are of special interest in the development of sepals, petals and ovules.

A total of three genes have been isolated from *Petunia hybrida* that are similar to *AP2*: *P. hybrida APETALA2A* (*PhAP2A*), *PhAP2B* and *PhAP2C*. *PhAP2A* is, to a large degree, homologous with the *AP2* gene of *Arabidopsis*, both in its sequence and in its expression pattern, which suggests that the two genes are orthologs. The proteins *PhAP2B* and *PhAP2C*, on the other hand, are slightly different, even though they belong to the family of transcription factors that are similar to *AP2*. In addition they are expressed in different ways, although they are very similar in comparison with *PhAP2A*. In fact, the mutants for these genes do not show the usual phenotype, that of the null alleles of A genes. A true A-function gene has not been found in Petunia; though a part of the A-function (the inhibition of the C in the outer two whorls) has been largely attributed to miRNA169 (colloquially called BLIND).

Genes Exhibiting type-B Function

In *A. thaliana* the type-B function mainly arises from two genes, *APETALA3* (*AP3*) and *PISTILLATA* (*PI*), both of which are MADS-box genes. A mutation of either of these genes causes the homeotic conversion of petals into sepals and of stamens into carpels. This also occurs in its orthologs in *A. majus*, which are DEFICIENS (*DEF*) and *GLOBOSA* (*GLO*) respectively. For both species the active form of binding with DNA is that derived from the heterodimer: AP3 and PI, or DEF and GLO, dimerize. This is the form in which they are able to function.

The *GLO/PI* lines that have been duplicated in *Petunia* contain *P. hybrida GLOBOSA1* (*PhGLO1*, also called *FBP1*) and also *PhGLO2* (also called *PMADS2* or *FBP3*). For the functional elements equivalent to *AP3/DEF* in *Petunia* there is both a gene that possesses a relatively similar sequence, called *PhDEF* and there is also an atypical B function gene called PhTM6. Phylogenetic studies have placed the first three within the «euAP3» lineage, while PhTM6 belongs to that of «paleoAP3». It is worth pointing out that, in terms of evolutionary history, the appearance of the euAP3 line seems to be related with the emergence of dicotyledons, as representatives of euAP3-type B function genes are present in dicotyledons while paleoAP3 genes are present in monocotyledons and basal angiosperms, among others.

As discussed above, the floral organs of eudicotyledonous angiosperms are arranged in 4 different verticils, containing the sepals, petals, stamen and carpels. The ABC model states that the identity of these organs is determined by the homeotic genes A, A+B, B+C and C, respectively. In contrast with the sepal and petal verticils of the eudicots, the perigone of many plants of the Liliaceae family have two nearly identical external petaloid verticils (the tepals). In order to explain the floral morphology of the Liliaceae, van Tunen *et al.* proposed a modified ABC model in 1993. This model suggests that class B genes are not only expressed in verticils 2 and 3, but also in 1. It therefore follows that the organs of verticils 1 and 2 express class A and B genes and this is how they have a petaloid structure. This theoretical model has been experimentally proven through the cloning and characterization of homologs of the *Antirrhinum* genes *GLOBOSA* and *DEFICIENS* in a Liliaceae, the tulip *Tulipa gesneriana*. These genes are expressed in verticils 1,2 and 3. The homologs *GLOBOSA* and *DEFICIENS* have also been isolated and characterized in *Agapanthus praecox* ssp. *orientalis* (Agapanthaceae), which is phylogenetically distant from the model organisms. In this study the genes were called *ApGLO* and *ApDEF*, respectively. Both contain open reading frames that code for proteins with 210 to 214 amino acids. Phylogenetic analysis of these sequences indicated that they belong to B gene family of the monocotyledons. In situ hybridization studies revealed that both sequences are expressed in verticil 1 as well as in 2 and 3. When taken together, these observations show that the floral development mechanism of *Agapanthus* also follows the modified ABC model.

Genes Exhibiting Type-C Function

In *A. thaliana*, the C function is derived from one MADS-box type gene called *AGAMOUS* (*AG*), which intervenes both in the establishment of stamen and carpel identity as well as in the determination of the floral meristem. Therefore, the *AG* mutants are devoid of androecium and gynoecium and they have petals and sepals in their place. In addition, the growth in the centre of the flower is undifferentiated, therefore the petals and sepals grow in repetitive verticils.

The *PLENA* (*PLE*) gene is present in *A. majus*, in place of the *AG* gene, although it is not an ortholog. However, the *FARINELLI* (*FAR*) gene is an ortholog, which is specific to the development of the anthers and the maturation of pollen.

In *Petunia*, *Antirrhinum* and in maize the C function is controlled by a number of genes that act in the same manner. The genes that are closer homologs of *AG* in *Petunia* are *pMADS3* and *floral-binding protein 6* (*FBP6*).

Genes Exhibiting Type-D and E Functions

The D function genes were discovered in 1995. These genes are MADS-box proteins and they have a function that is distinct from those previously described, although they have a certain homology with C function genes. These genes are called *FLORAL BINDING PROTEIN7* (*FBP7*) and *FLORAL BINDING PROTEIN1L* (*FBP1l*). It was found that, in *Petunia*, they are involved in the development of the ovule. Equivalent genes were later found in *Arabidopsis*, where they are also involved in controlling the development of carpels and the ovule and even with structures related to seed dispersal.

The appearance of interesting phenotypes in RNA interference studies in *Petunia* and tomato led, in 1994, to the definition of a new type of function in the floral development model. The E function was initially thought to be only involved in the development of the three innermost verticils, however, subsequent work found that its expression was required in all the floral verticils.

Inflorescence

Aloe hereroensis, showing inflorescence with branched peduncle

An inflorescence is a group or cluster of flowers arranged on a stem that is composed of a main branch or a complicated arrangement of branches. Morphologically, it is the modified part of the shoot of seed plants where flowers are formed. The modifications can involve the length and the nature of the internodes and the phyllotaxis, as well as variations in the proportions, compressions, swellings, adnations, connations and reduction of main and secondary axes. Inflorescence can also be defined as the reproductive portion of a plant that bears a cluster of flowers in a specific pattern.

The stem holding the whole inflorescence is called a peduncle and the major axis (incorrectly referred to as the main stem) holding the flowers or more branches within the inflorescence is called the rachis. The stalk of each single flower is called a pedicel. A flower that is not part of an inflorescence is called a solitary flower and its stalk is also referred to as a peduncle. Any flower in an inflorescence may be referred to as a floret, especially when the individual flowers are particularly small and borne in a tight cluster, such as in a pseudanthium. The fruiting stage of an inflorescence is known as an infructescence.

Inflorescences may be simple (single) or complex (panicle). The rachis may be one of several types, including Single, Composite, Umbel, Spike or Raceme.

General Characteristics

Inflorescences are described by many different characteristics including how the flowers are arranged on the peduncle, the blooming order of the flowers and how different clusters of flowers are grouped within it. These terms are general representations as plants in nature can have a combination of types.

Bracts

Inflorescences usually have modified shoots foliage different from the vegetative part of the plant. Considering the broadest meaning of the term, any leaf associated with an inflorescence is called a bract. A bract is usually located at the node where the main stem of the inflorescence forms, joined to the main stem of the plant, but other bracts can exist within the inflorescence itself. They serve a variety of functions which include attracting pollinators and protecting young flowers. According to the presence or absence of bracts and their characteristics we can distinguish:

- Ebracteate inflorescences: No bracts in the inflorescence.
- Bracteate inflorescences: The bracts in the inflorescence are very specialised, sometimes reduced to small scales, divided or dissected.
- Leafy inflorescences: Though often reduced in size, the bracts are unspecialised and look like the typical leaves of the plant, so that the term **flowering stem** is usually applied instead of inflorescence. This use is not technically correct, as, despite their 'normal' appearance, these *leaves* are considered, in fact, *bracts*, so that 'leafy inflorescence' is preferable.
- Leafy-bracted inflorescences: Intermediate between bracteate and leafy inflorescence.

If many bracts are present and they are strictly connected to the stem, like in the family Asteraceae, the bracts might collectively be called an involucre. If the inflorescence has a second unit of bracts further up the stem, they might be called an involucel.

Ebracteate inflorescence.

Ebracteate of *Wisteria sinensis*

Bracteate inflorescence.

Bracteate inflorescence of *Pedicularis verticillata*.

Leafy-bracted inflorescence.

Leafy-bracted inflorescence of *Rhinanthus angustifolius*.

Leafy inflorescence.

Leafy inflorescence of *Aristolochia clematitis*.

Terminal Flower

Plant organs can grow according to two different schemes, namely monopodial or racemose and sympodial or cymose. In inflorescences these two different growth patterns are called indeterminate and determinate respectively, and indicate whether a terminal flower is formed and where flowering starts within the inflorescence.

- Indeterminate inflorescence: Monopodial (racemose) growth. The terminal bud keeps growing and forming lateral flowers. A terminal flower is never formed.
- Determinate inflorescence: Sympodial (cymose) growth. The terminal bud forms a terminal flower and then dies out. Other flowers then grow from lateral buds.

Indeterminate and determinate inflorescences are sometimes referred to as open and closed inflorescences respectively.

In an indeterminate inflorescence there is no true terminal flower and the stem usually has a rudimentary end. In many cases the last true flower formed by the terminal bud (subterminal flower) straightens up, appearing to be a terminal flower. Often a vestige of the terminal bud may be noticed higher on the stem.

In determinate inflorescences the terminal flower is usually the first to mature (precursive development), while the others tend to mature starting from the bottom of the stem. This pattern is called acropetal maturation. When flowers start to mature from the top of the stem, maturation is basipetal, while when the central mature first, divergent.

Indeterminate inflorescence with a perfect acropetal maturation.

Indeterminate inflorescence with an acropetal maturation and lateral flower buds.

Indeterminate inflorescence with the subterminal flower to simulate the terminal one (vestige present)

Determinate inflorescence with acropetal maturation

Determinate inflorescence with basipetal maturation

Determinate inflorescence with divergent maturation

Phyllotaxis

As with leaves, flowers can be arranged on the stem according to many different patterns. See 'Phyllotaxis' for in-depth descriptions

Alternate flowers

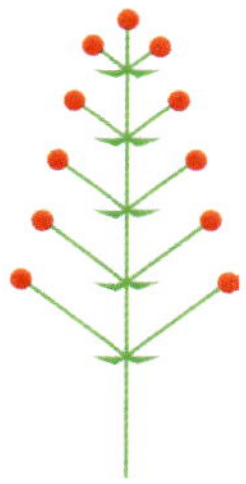

Opposite flowers

Similarly arrangement of leaf in bud is called Ptyxis.

Metatopy

Metatopy is the placement of organs out of their normally expected position: typically metatopy occurs in inflorescences when unequal growth rates alter different areas of the axis and the organs attached to the axis.

When a single or a cluster of flower(s) is located at the axil of a bract, the location of the bract in relation to the stem holding the flower(s) is indicated by the use of different terms and may be a useful diagnostic indicator.

Typical placement of bracts include:

- Some plants have bracts that subtend the inflorescence, where the flowers are on branched stalks; the bracts are not connected to the stalks holding the flowers, but are adnate or attached to the main stem (Adnate describes the fusing together of different unrelated parts. When the parts fused together are the same, they are connately joined.)
- Other plants have the bracts subtend the pedicel or peduncle of single flowers.

Metatopic placement of bracts include:

- When the bract is attached to the stem holding the flower (the pedicel or peduncle), it is said to be recaulescent; sometimes these bracts or bracteoles are highly modified and appear to be appendages of the flower calyx. Recaulescences is the fusion of the subtending leaf with the stem holding the bud or the bud itself, thus the leaf or bract is adnate to the stem of flower.
- When the formation of the bud is shifted up the stem distinctly above the subtending leaf, it is described as concaulescent.

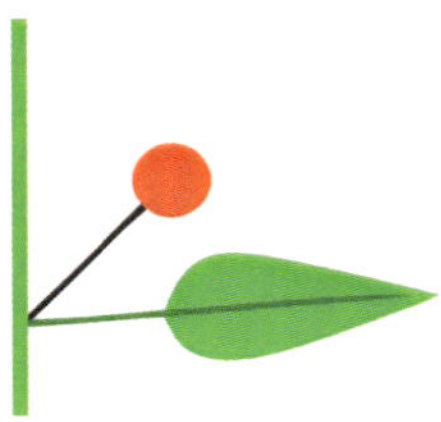

Flower and subtending bract

Lilium martagon (flower and subtending bract)

Concaulescence

Solanum lycopersicum (concaulescence)

Recaulescence

Tilia cordata (recaulescence)

Organization

There is no general consensus in defining the different inflorescences. The following is based on Focko Weberling's *Morphologie der Blüten und der Blütenstände* (Stuttgart, 1981). The main groups of inflorescences are distinguished by branching. Within these groups, the most important characteristics are the intersection of the axes and different variations of the model. They may contain many flowers (pluriflor) or a few (pauciflor). Inflorescences can be simple or compound.

Simple Inflorescences

Inflorescence of sessile disc florets forming the capitulum

Indeterminate or Racemose

Indeterminate simple inflorescences are generally called racemose. The main kind of racemose inflorescence is the raceme, from classical Latin *racemus*, cluster of grapes. The other kind of racemose inflorescences can all be derived from this one by dilation, compression, swelling or reduction of the different axes. Some passage forms between the obvious ones are commonly admitted.

- A raceme is an unbranched, indeterminate inflorescence with pedicellate (having short floral stalks) flowers along the axis.
- A spike is a type of raceme with flowers that do not have a pedicel.
- A racemose corymb is an unbranched, indeterminate inflorescence that is flat-topped or convex due to their outer pedicels which are progressively longer than inner ones.
- An umbel is a type of raceme with a short axis and multiple floral pedicels of equal length that appear to arise from a common point.
- A spadix is a spike of flowers densely arranged around it, enclosed or accompanied by a highly specialised bract called a spathe. It is characteristic of the Araceae family.
- A flower head or capitulum is a very contracted raceme in which the single sessile flowers share are borne on an enlarged stem. It is characteristic of Dipsacaceae.
- A catkin or ament is a scaly, generally drooping spike or raceme. Cymose or other complex inflorescences that are superficially similar are also generally called thus.

Raceme

Epilobium angustifolium

Spike

Plantago media (spike)

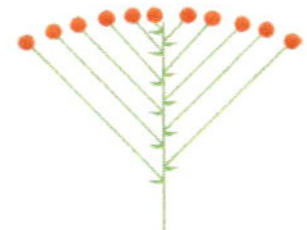
Racemose corymb

Iberis umbellata (racemose corymb)

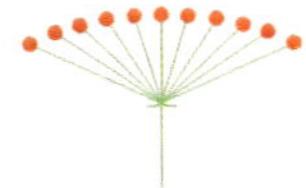
Umbel

Astrantia minor (umbel)

Spadix

Catkin (racemose or spicate)

Arum maculatum (spadix)

Head (round)

Dipsacus fullonum (head)

Alnus incana (ament)

Determinate or Cymose

Determinate simple inflorescences are generally called cymose. The main kind of cymose inflorescence is the cyme (pronounced 'saim', from the Latin *cyma* in the sense 'cabbage sprout', from Greek *kuma* 'anything swollen'). Cymes are further divided according to this scheme:

- Only one secondary axis: monochasium
  - Secondary buds always develop on the same side of the stem: helicoid cyme or bostryx
    - The successive pedicels are aligned on the same plane: drepanium
  - Secondary buds develop alternately on the stem : scorpioid cyme
    - The successive pedicels are arranged in a sort of spiral: cincinnus (characteristic of the Boraginaceae and Commelinaceae)
    - The successive pedicels follow a zig-zag path on the same plane: rhipidium (many Iridaceae)
- Two secondary axes: dichasial cyme
  - Secondary axis still dichasial: dichasium (characteristic of Caryophyllaceae)
  - Secondary axis monochasia: double scorpioid cyme or double helicoid cyme
- More than two secondary axes: pleiochasium

Double cyme

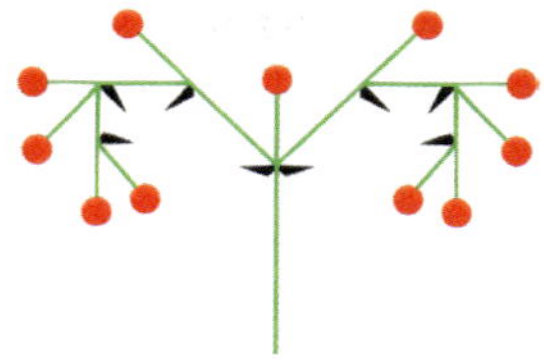

Double cyme

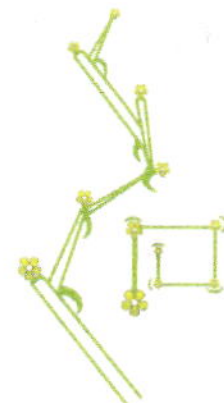

Bostryx (lateral and top view)

Hypericum perforatum (bostryx)

Drepanium (lateral and top view)

Gladiolus imbricatus (drepanium)

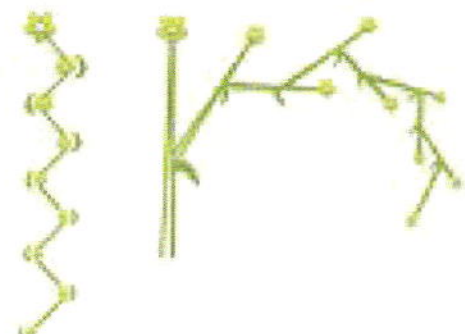

Cincinnus (lateral and top view)

Symphytum officinale (cincinnus)

Rhipidium (lateral and top view)

Canna sp. (rhipidium)

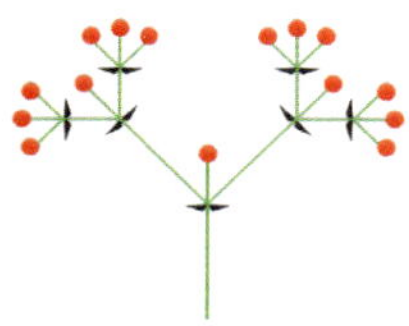

Dichasium

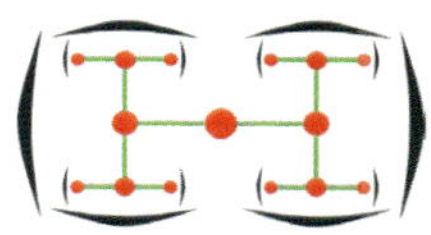

Dichasium, top view

Silene dioica (dichasium)

A cyme can also be so compressed that it looks like an umbel. Strictly speaking this kind of inflorescence could be called umbelliform cyme, although it is normally called simply 'umbel'.

Another kind of definite simple inflorescence is the raceme-like cyme or botryoid; that is as a raceme with a terminal flower and is usually improperly called 'raceme'.

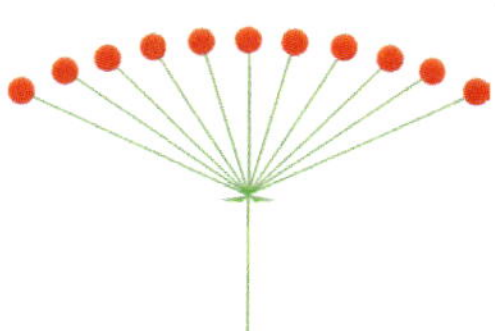

Umbelliform cyme

Pelargonium zonale (umbelliform cyme)

Botryoid

Berberis vernae (botryoid)

A reduced raceme or cyme that grows in the axil of a bract is called a fascicle. A verticillaster is a fascicle with the structure of a dichasium; it is common among the Lamiaceae. Many verticillasters with reduced bracts can form a spicate (spike-like) inflorescence that is commonly called a spike.

Gentiana lutea (fascicles)

Lamium orvala (verticillaster)

Mentha longifolia ('spike')

Compound Inflorescences

Simple inflorescences are the basis for compound inflorescences or synflorescences. The single flowers are there replaced by a simple inflorescence, which can be both a racemose or a cymose one. Compound inflorescences are composed of branched stems and can involve complicated arrangements that are difficult to trace back to the main branch.

A kind of compound inflorescence is the double inflorescence, in which the basic structure is repeated in the place of single florets. For example a double raceme is a raceme in which the single flowers are replaced by other simple racemes; the same structure can be repeated to form triple or more complex structures.

Compound raceme inflorescences can either end with a final raceme (homoeothetic), or not (heterothetic). A compound raceme is often called a panicle. Note that this definition is very different from that given by Weberling.

Compound umbels are umbels in which the single flowers are replaced by many smaller umbels called umbellets. The stem attaching the side umbellets to the main stem is called a ray.

Homeothetic compound raceme

Melilotus officinalis (homoeothetic compound raceme)

Heterothetic compound raceme

Hebe albicans (heterothetic compound raceme)

Compound spike

Lolium temulentum (compound spike)

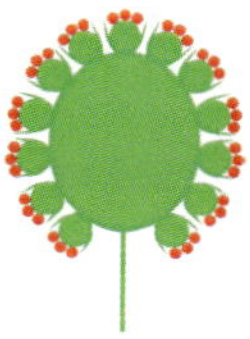

Compound capitulum

Echinops ritro (compound capitulum)

Compound (double) umbel

Laserpicium latifolium (double umbel)

Compound (triple) umbel

The most common kind of definite compound inflorescence is the panicle (of Webeling, or 'panicle-like cyme'). A panicle is a definite inflorescence that is increasingly more strongly and irregularly branched from the top to the bottom and where each branching has a terminal flower.

The so-called cymose corymb is similar to a racemose corymb but has a panicle-like structure. Another type of panicle is the anthela. An anthela is a cymose corymb with the lateral flowers higher than the central ones.

Panicle

Vitis vinifera (panicle)

Cymose corymb

Sambucus nigra (cymose corymb)

Anthela

Juncus inflexus (anthela)

A raceme in which the single flowers are replaced by cymes is called a (indefinite) thyrse. The secondary cymes can be of any of the different types of dichasia and monochasia. A botryoid in which the single flowers are replaced by cymes is a definite thyrse or thyrsoid. Thyrses are often confusingly called panicles.

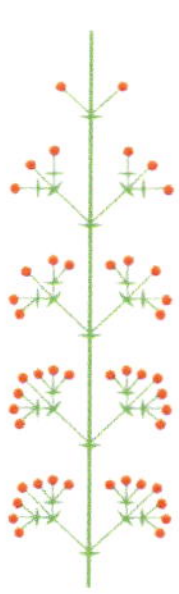

Thyrse

Aesculus hippocastanum

Thyrsoid

Syringa vulgaris

Other combinations are possible. For example, heads or umbels may be arranged in a corymb or a panicle.

Achillea sp. (heads in a corymb)

Hedera helix (umbels in a panicle)

Other

The family Asteraceae is characterised by a highly specialised head technically called a calathid (but usually referred to as 'capitulum' or 'head'). The family Poaceae has a peculiar inflorescence of small spikes (spikelets) organised in panicles or spikes that are usually simply and improperly referred to as spike and panicle. The genus *Ficus* (Moraceae) has an inflorescence called syconium and the genus *Euphorbia* has **cy**athia (sing. *cyathium*), usually organised in umbels.

Matricaria chamomilla (calathid)

Triticum aestivum (compound spikes, "spikes")

Oryza sativa (spikes in a panicle, "panicle")

Ficus carica (syconium)

Euphorbia tridentata (cyathium)

Euphorbia cyparissias (cyathia in an umbel)

(*Coleus*-false spike)

Development and Patterning

Genetic Basis

Genes that shape inflorescence development have been studied at great length in Arabidopsis. LEAFY (LFY) is a gene that promotes floral meristem identity, regulating inflorescence development in Arabidopsis. Any alterations in timing of LFY expression can cause formation of different inflorescences in the plant. Genes similar in function to LFY include APETALA1 (AP1). Mutations in LFY, AP1, and similar promoting genes can cause conversion of flowers into shoots. In contrast to LEAFY, genes like terminal flower (TFL) support the activity of an inhibitor that prevents flowers from growing on the inflorescence apex (flower primordium initiation), maintaining inflorescence meristem identity. Both types of genes help shape flower development in accordance with the ABC model of flower development. Studies have been recently conducted or are ongoing for homologs of these genes in other flower species.

Environmental Influences

Inflorescence-feeding insect herbivores shape inflorescences by reducing lifetime fitness (how much flowering occurs), seed production by the inflorescences, and plant density, among other traits. In the absence of this herbivory, inflorescences usually produce more flower heads and seeds. Temperature can also variably shape inflorescence development. High temperatures can impair the proper development of flower buds or delay bud development in certain species, while in others, an increase in temperature can hasten inflorescence development.

Meristems and Inflorescence Architecture

The shift from the vegetative to reproductive phase of a flower involves the development of an inflorescence meristem that generates floral meristems. Plant inflorescence architecture depends on which meristems becomes flowers and which become shoots. Consequently, genes that regulate floral meristem identity play major roles in determining inflorescence architecture because their expression domain will direct where the plant's flowers are formed.

On a larger scale, inflorescence architecture affects the quality and quantity of offspring from selfing and outcrossing, as the architecture can influence pollination success. For example, Asclepias inflorescences have been shown to have an upper size limit, shaped by self-pollination levels due to crosses between inflorescences on the same plant or between flowers on the same inflorescence. In Aesculus sylvatica, it has been shown that the most common inflorescence sizes are correlated with the highest fruit production as well.

Issues with Inflorescence Identification

Some species have flower and inflorescence intermediates. In these cases, some reproductive structures of certain flowers appear as transitional between inflorescences and flowers, making it difficult to accurately categorize and identify the structure as one or the other. For example, the genus Potamogeton of the family Potamogetonaceae are considered to have inflorescences that appear like an individual flower.

Perianth

The perianth (perigonium, perigon or perigone) is the non-reproductive part of the flower, and structure that forms an envelope surrounding the sexual organs, consisting of the calyx (sepals) and the corolla (petals). The term perianth is derived from the Greek *peri*, meaning around, and *anthos*, meaning flower, while perigonium is derived from *gonos*, meaning seed, i.e. sexual organs. In the mosses and liverworts (Marchantiophyta), the perianth is the sterile tubelike tissue that surrounds the female reproductive structure (or developing sporophyte).

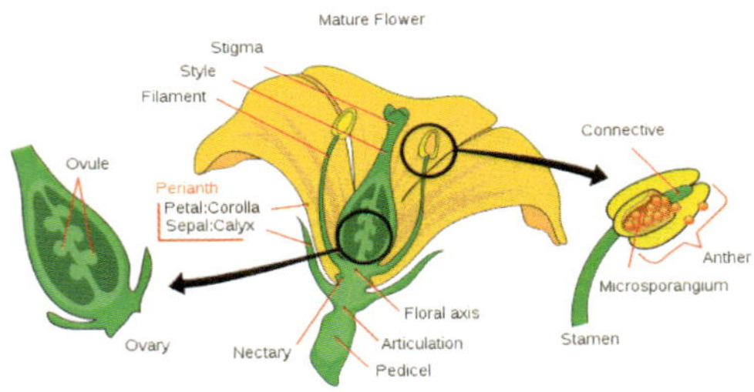

Diagram showing the parts of a mature flower. In this example the perianth is separated into a calyx (sepals) and corolla (petals)

Flowering Plants

In flowering plants, the perianth may be described as being either dichlamydeous/heterochlamydeous in which the calyx and corolla are clearly separate, or homochlamydeous, in which they are indistinguishable (and the sepals and petals are collectively referred to as tepals). When the perianth is in two whorls, it is described as biseriate. While the calyx may be green, it may also be brightly coloured, and is then described as petaloid. When the undifferentiated tepals resemble petals, they are also referred to as "petaloid", as in petaloid monocots, orders of monocots with brightly coloured tepals. Since they include Liliales, an alternative name is lilioid monocots. The corolla and petals have a role in attracting pollinators, but this may be augmented by more specialised structures like the corona.

When the corolla consists of separate tepals the term apotepalous is used, or syntepalous if the tepals are fused to one another. The petals may be united to form a tubular corolla (gamopetalous or sympetalous). If either the petals or sepals are entirely absent, the perianth can be described as being monochlamydeous.

Types of perianth

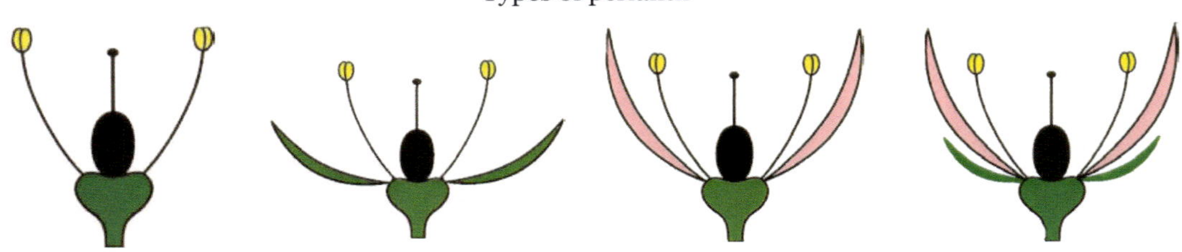

Achlamydeous floral meristem without a corolla or calyx

Monochlamydeous perianth with non-petaloid calyx only

Monochlamydeous perianth with corolla only or homochlamydeous perigonium with tepals

Dichlamydeous/heterochlamydeous perianth with separate whorls

Both sepals and petals may have stomata and veins, even if vestigial. In some taxa, for instance some magnolias and water lilies the perianth is arranged in a spiral on nodes, rather than whorls. Flowers with spiral perianths tend to also be those with undifferentiated perianths.

Corona

Flower of *Narcissus* showing an outer white corolla with a central yellow corona (paraperigonium)

Flower of *Passiflora incarnata* showing corona of fine appendages between petals and stamens

An additional structure in some plants (e.g. *Narcissus*, *Passiflora* (passion flower), some *Hippeastrum*, Liliaceae) is the corona (paraperigonium, paraperigon, or paracorolla), a ring or set of appendages of adaxial tissue arising from the corolla or the outer edge of the stamens. It is often positioned where the corolla lobes arise from the corolla tube.

The pappus of Asteraceae, considered to be a modified calyx, is also called a corona if it is shaped like a crown.

Floral Symmetry

[Left] Normal *Streptocarpus* flower (zygomorphic or mirror-symmetric), and [right] peloric (radially symmetric) flower on the same plant

Floral symmetry describes whether, and how, a flower, in particular its perianth, can be divided into two or more identical or mirror-image parts.

Uncommonly, flowers may have no axis of symmetry at all, typically because their parts are spirally arranged.

Actinomorphic

Wurmbea stricta, its tepals in actinomorphic arrangement

Most flowers are actinomorphic ("star shaped", "radial"), meaning they can be divided into 3 or more identical sectors which are related to each other by rotation about the centre of the flower. Typically, each sector might contain one tepal or one petal and one sepal and so on. It may or may not be possible to divide the flower into symmetrical halves by the same number of longitudinal planes passing through the axis: Oleander is an example of a flower without such mirror planes. Actinomorphic flowers are also called radially symmetrical or regular flowers. Other examples of actinomorphic flowers are the lily (*Lilium*, Liliaceae) and the buttercup (*Ranunculus*, Ranunculaceae).

Zygomorphic

Satyrium carneum. Ground orchid with typical zygomorphic floral anatomy

Zygomorphic flowers can be divided by only a single plane into two mirror-image halves, much like a yoke or a person's face. Examples are orchids and the flowers of most members of the Lamiales (e.g., Scrophulariaceae and Gesneriaceae). Some authors prefer the term monosymmetry or bilateral symmetry.

Asymmetry

A few plant species have flowers lacking any symmetry, and therefore having a "handedness". Examples are *Valeriana officinalis* and *Canna indica*.

Differences

Actinomorphic flowers are a basal angiosperm character; zygomorphic flowers are a derived character that has evolved many times.

Some familiar and seemingly actinomorphic so-called flowers, such as those of daisies and dandelions (Asteraceae), and most species of *Protea*, are actually clusters of tiny (not necessarily actinomorphic) flowers arranged into a roughly radially symmetric inflorescence of the form known as a head, capitulum, or pseudanthium.

Peloria

Digitalis purpurea (common foxglove) displaying an aberrant peloric terminal flower and normal zygomorphic flowers

Peloria or a peloric flower is the aberration in which a plant that normally produces zygomorphic flowers produces actinomorphic flowers instead. This aberration can be developmental, or it can have a genetic basis: the CYCLOIDEA gene controls floral symmetry. Peloric *Antirrhinum* plants have been produced by knocking out this gene. Many modern cultivars of *Sinningia speciosa* ("gloxinia") have been bred to have peloric flowers as they are larger and showier than the normally zygomorphic flowers of this species.

Charles Darwin explored peloria in *Antirrhinum* (snapdragon) while researching the inheritance of floral characteristics for his *The Variation of Animals and Plants under Domestication*. Later research, using *Digitalis purpurea*, showed that his results were largely in line with Mendelian theory.

Symmetry Groups

If we consider only those flowers which consist in a single flower, rather than a flower head or inflorescence, we can group the flowers into a relatively small number of 2D symmetry groups. Monocots are identifiable by their trimerous petals, thus monocots often have rotational symmetry of order 3. If the flower also has 3 lines of mirror symmetry the group it belongs to is the dihedral group D3. If not, then it belongs to the cyclic group C3. Eudicots with tetramerous or pentamerous petals may have rotational symmetry of order 4 or 5. Again, whether they also have mirror planes decides whether they belong to dihedral (D4 and D5) or cyclic groups (C4 or C5). The sepals of some monocot flowers develop to replicate the petals, thus, superficially, certain monocots can appear to have rotational symmetry of order 6 and belong to either symmetry group D6 or C6. However, flower symmetry is rarely perfect.

Petal

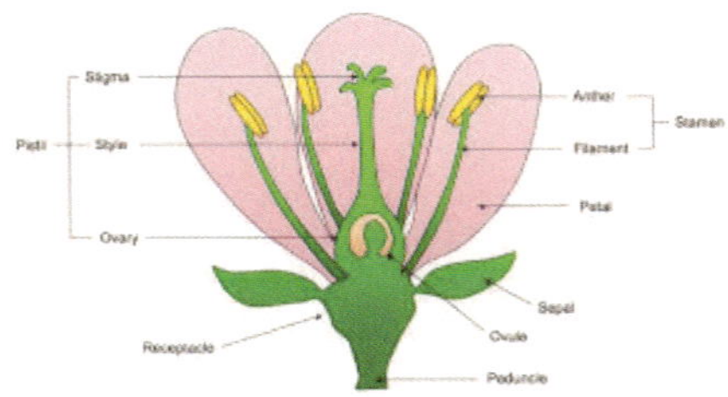

Diagram showing the parts of a mature flower. In this example the perianth is separated into a calyx (sepals) and corolla (petals)

Petals are modified leaves that surround the reproductive parts of flowers. They are often brightly colored or unusually shaped to attract pollinators. Together, all of the petals of a flower are called a *corolla*. Petals are usually accompanied by another set of special leaves called sepals, that collectively form the *calyx* and lie just beneath the corolla. The calyx and the corolla together make up the perianth. When the petals and sepals of a flower are difficult to distinguish, they are collectively called tepals. Examples of plants in which the term *tepal* is appropriate include genera such as *Aloe* and *Tulipa*. Conversely, genera such as *Rosa* and *Phaseolus* have well-distinguished sepals and petals. When the undifferentiated tepals resemble petals, they are referred to as "petaloid", as in petaloid monocots, orders of monocots with brightly coloured tepals. Since they include Liliales, an alternative name is lilioid monocots.

Although petals are usually the most conspicuous parts of animal-pollinated flowers, wind-pollinated species, such as the grasses, either have very small petals or lack them entirely.

Tetrameric flower of a Primrose willowherb (*Ludwigia octovalvis*) showing petals and sepals

Corolla

Corolla forming a tube with long points (and a separate green calyx tube)

The role of the corolla in plant evolution has been studied extensively since Charles Darwin postulated a theory of the origin of elongated corollae and corolla tubes.

A corolla of separate tepals is *apopetalous*. If the petals are free from one another in the corolla, the plant is *polypetalous* or *choripetalous*; while if the petals are at least partially fused together, it is *gamopetalous* or *sympetalous*. In the case of fused tepals, the term is *syntepalous*. The corolla in some plants forms a tube.

Variations

Pelargonium peltatum: its floral structure is almost identical to that of geraniums, but it is conspicuously zygomorphic

Geranium incanum, with an actinomorphic flower typical of the genus

White pea, a zygomorphic flower

Narcissus pseudonarcissus, showing from the bend to the tip of the flower: spathe, floral cup, tepals, corona

Petals can differ dramatically in different species. The number of petals in a flower may hold clues to a plant's classification. For example, flowers on eudicots (the largest group of dicots) most frequently have four or five petals while flowers on monocots have three or six petals, although there are many exceptions to this rule.

The petal whorl or corolla may be either radially or bilaterally symmetrical. If all of the petals are essentially identical in size and shape, the flower is said to be regular or actinomorphic (meaning "ray-formed"). Many flowers are symmetrical in only one plane (i.e., symmetry is bilateral) and are termed irregular or zygomorphic (meaning "yoke-" or "pair-formed"). In irregular flowers, other floral parts may be modified from the *regular* form, but the petals show the greatest deviation from radial symmetry. Examples of zygomorphic flowers may be seen in orchids and members of the pea family.

In many plants of the aster family such as the sunflower, *Helianthus annuus*, the circumference of the flower head is composed of ray florets. Each ray floret is anatomically an individual flower with a single large petal. Florets in the centre of the disc typically have no or very reduced petals. In some plants such as *Narcissus* the lower part of the petals or tepals are fused to form a floral cup (hypanthium) above the ovary, and from which the petals proper extend.

Petal often consists of two parts: the upper, broad part, similar to leaf blade, also called the *blade* and the lower part, narrow, similar to leaf petiole, called the *claw*, separated from each other at the *limb*. Claws are developed in petals of some flowers of the family *Brassicaceae*, such as *Erysimum cheiri*.

The inception and further development of petals shows a great variety of patterns. Petals of different species of plants vary greatly in colour or colour pattern, both in visible light and in ultraviolet. Such patterns often function as guides to pollinators, and are variously known as nectar guides, pollen guides, and floral guides.

Genetics

The genetics behind the formation of petals, in accordance with the ABC model of flower development, are that sepals, petals, stamens, and carpels are modified versions of each other. It appears that the mechanisms to form petals evolved very few times (perhaps only once), rather than evolving repeatedly from stamens.

Significance of Pollination

Pollination is an important step in the sexual reproduction of higher plants. Pollen is produced by the male flower or by the male organs of hermaphroditic flowers.

Pollen does not move on its own and thus requires wind or animal pollinators to disperse the pollen to the stigma (botany) of the same or nearby flowers. However, pollinators are rather selective in determining the flowers they choose to pollinate. This develops competition between flowers and as a result flowers must provide incentives to appeal to pollinators (unless the flower self-pollinates or is involved in wind pollination). Petals play a major role in competing to attract pollinators. Henceforth pollination dispersal could occur and the survival of many species of flowers could prolong.

Functions and Purposes

Petals have various functions and purposes depending on the type of plant. In general, petals operate to protect some parts of the flower and attract/repel specific pollinators.

Function

This is where the positioning of the flower petals are located on the flower is the corolla e.g. the buttercup having shiny yellow flower petals which contain guidelines amongst the petals in aiding the pollinator towards the nectar. Pollinators have the ability to determine specific flowers they wish to pollinate. Using incentives flowers draw pollinators and set up a mutual relation between each other in which case the pollinators will remember to always guard and pollinate these flowers (unless incentives are not consistently met and competition prevails).

Scent

The petals could produce different scents to allure desirable pollinators or repel undesirable pollinators. Some flowers will also mimic the scents produced by materials such as decaying meat, to attract pollinators to them.

Colour

Various colour traits are used by different petals that could attract pollinators that have poor smelling abilities, or that only come out at certain parts of the day. Some flowers are able to change the colour of their petals as a signal to mutual pollinators to approach or keep away.

Shape and Size

Furthermore, the shape and size of the flower/petals is important in selecting the type of pollinators they need. For example, large petals and flowers will attract pollinators at a large distance or that are large themselves. Collectively the scent, colour and shape of petals all play a role in attracting/repelling specific pollinators and providing suitable conditions for pollinating. Some pollinators include insects, birds, bats and the wind. In some petals, a distinction can be made between a lower narrowed, stalk-like basal part referred to as the claw, and a wider distal part referred to as the blade. Often the claw and blade are at an angle with one another.

Types of Pollination

Wind Pollination

Wind-pollinated flowers often have small, dull petals and produce little or no scent. Some of these flowers will often have no petals at all. Flowers that depend on wind pollination will produce large amounts of pollen because most of the pollen scattered by the wind tends to not reach other flowers.

Attracting Insects

Flowers have various regulatory mechanisms in order to attract insects. One such helpful mechanism is the use of colour guiding marks. Insects such as the bee or butterfly can see the ultraviolet marks which are contained on these flowers, acting as an attractive mechanism which is not visible towards the human eye. Many flowers contain a variety of shapes acting to aid with the landing of the visiting insect and also influence the insect to brush against anthers and stigmas (parts of the flower). One such example of a flower is the pōhutukawa (*Metrosideros excelsa*) which acts to regulate colour within a different way. The pōhutukawa contains small petals also having bright large red clusters of stamens. Another attractive mechanism for flowers is the use of scent which is highly attractive towards humans such as the rose, but some are very fragrant within attracting flies as they have a smell of rotting meat. Dark is another factor in which flowers have grown to adapt these conditions so colour lacks vision at night therefore scent is the solution for flowers which are pollinated by night flying insects such as the moth.

Attracting Birds

Flowers are also pollinated by birds and must be large and colorful to be visible against natural scenery. Such bird –pollinated native plants include: Kōwhai (*Sophora* species), flax (*Phormium tenax*, harakeke) and kākā beak (*Clianthus puniceus*, kōwhai ngutu-kākā). Interestingly enough, flowers adapt the mechanism on their petals to change colour in acting as a communicative mechanism for the bird to visit. An example is the tree fuchsia (*Fuchsia excorticata*, kōtukutuku) which are green when needing to be pollinated and turn red for the birds to stop coming and pollinating the flower.

Bat-pollinated Flowers

Flowers can be pollinated by short tailed bats. An example of this is the dactylanthus (*Dactylanthus taylorii*). This plant has its home under the ground acting the role of a parasite on the roots of forest trees. The dactylanthus has only its flowers pointing to the surface and the flowers lack colour but have the advantage of containing lots of nectar and a very strong scent. These act as a very useful mechanism in attracting the bat.

Sepal

Tetramerous flower of *Ludwigia octovalvis* showing petals and sepals.

After blooming, the sepals of *Hibiscus sabdariffa* expand into an edible accessory fruit

In many Fabaceae flowers, a calyx tube surrounds the petals.

A sepal is a part of the flower of angiosperms (flowering plants). Usually green, sepals typically function as protection for the flower in bud, and often as support for the petals when in bloom. The term *sepalum* was coined by Noël Martin Joseph de Necker in 1790, and derived from the Greek (*skepi*), a covering.

Collectively the sepals are called the calyx (plural calyces), the outermost whorl of parts that form a flower. The word *calyx* was adopted from the Latin , not to be confused with *calix*, a cup or goblet. *Calyx* derived from the Greek , a bud, a calyx, a husk or wrapping, (cf Sanskrit *kalika*, a bud) while *calix* derived from the Greek , a cup or goblet, and the words have been used interchangeably in botanical Latin.

After flowering, most plants have no more use for the calyx which withers or becomes vestigial. Some plants retain a thorny calyx, either dried or live, as protection for the fruit or seeds. Examples include species of *Acaena*, some of the Solanaceae (for example the Tomatillo, *Physalis philadelphica*), and the water caltrop, *Trapa natans*. In some species the calyx not only persists after flowering, but instead of withering, begins to grow until it forms a bladder-like enclosure around the fruit. This is an effective protection against some kinds of birds and insects, for example in *Hibiscus trionum* and the Cape gooseberry.

Morphologically, both sepals and petals are modified leaves. The calyx (the sepals) and the corolla (the petals) are the outer sterile whorls of the flower, which together form what is known as the *perianth*.

The term *tepal* is usually applied when the parts of the perianth are difficult to distinguish, e.g. the petals and sepals share the same color, or the petals are absent and the sepals are colorful. When the undifferentiated tepals resemble petals, they are referred to as "petaloid", as in petaloid monocots, orders of monocots with brightly coloured tepals. Since they include Liliales, an alternative name is lilioid monocots. Examples of plants in which the term tepal is appropriate include genera such as *Aloe* and *Tulipa*. In contrast, genera such as *Rosa* and *Phaseolus* have well-distinguished sepals and petals.

The number of sepals in a flower is its merosity. Flower merosity is indicative of a plant's classification. The merosity of a eudicot flower is typically four or five. The merosity of a monocot or palaeodicot flower is three, or a multiple of three.

The development and form of the sepals vary considerably among flowering plants. They may be free (polysepalous) or fused together (gamosepalous). Often, the sepals are much reduced, appearing somewhat awn-like, or as scales, teeth, or ridges. Most often such structures protrude until the fruit is mature and falls off.

Examples of flowers with much reduced perianths are found among the grasses.

In some flowers, the sepals are fused towards the base, forming a calyx tube (as in the Lythraceae family, and Fabaceae). In other flowers (e.g., Rosaceae, Myrtaceae) a hypanthium includes the bases of sepals, petals, and the attachment points of the stamens.

Plant Reproductive Morphology

Close-up of a flower of *Schlumbergera* (Christmas or Holiday Cactus), showing part of the gynoecium (the stigma and part of the style is visible) and the stamens that surround it

Plant reproductive morphology is the study of the physical form and structure (the morphology) of those parts of plants directly or indirectly concerned with sexual reproduction.

Among all living organisms, flowers, which are the reproductive structures of angiosperms, are the most varied physically and show a correspondingly great diversity in methods of reproduction. Plants that are not flowering plants (green algae, mosses, liverworts, hornworts, ferns and gymnosperms such as conifers) also have complex interplays between morphological adaptation and environmental factors in their sexual reproduction. The breeding system, or how the sperm from one plant fertilizes the ovum of another, depends on the reproductive morphology, and is the single most important determinant of the genetic structure of nonclonal plant populations. Christian Konrad Sprengel (1793) studied the reproduction of flowering plants and for the first time it was understood that the pollination process involved both biotic and abiotic interactions. Charles

Darwin's theories of natural selection utilized this work to build his theory of evolution, which includes analysis of the coevolution of flowers and their insect pollinators.

Use of Sexual Terminology

Dioicous gametophytes of the liverwort *Marchantia polymorpha.*

In this species, gametes are produced on different plants on umbrella-shaped gametophores with different morphologies. The radiating arms of female gameteophores (left) protect archegonia that produce eggs. Male gametophores (right) are topped with antheridia that produce sperm.

Plants have complex lifecycles involving alternation of generations. One generation, the sporophyte, gives rise to the next generation asexually via spores. Spores may be identical isospores or come in different sizes (microspores and megaspores), but strictly speaking, spores and sporophytes are neither male nor female because they do not produce gametes. The alternate generation, the gametophyte, produces gametes, eggs and/or sperm. A gametophyte can be monoicous (bisexual), producing both eggs and sperm or dioicous (unisexual), either female (producing eggs) or male (producing sperm).

In the bryophytes (liverworts, mosses and hornworts), the sexual gametophyte is the dominant generation. In ferns and seed plants (including cycads, conifers, flowering plants, etc.) the sporophyte is the dominant generation. The obvious visible plant, whether a small herb or a large tree, is the sporophyte, and the gametophyte is very small. In seed plants, each female gametophyte, and the spore that gives rise to it, is hidden within the sporophyte and is entirely dependent on it for nutrition. Each male gametophyte typically consists of from two to four cells enclosed within the protective wall of a pollen grain.

The sporophyte of a flowering plant is often described using sexual terms (e.g. "female" or "male") based on the sexuality of the gametophyte it gives rise to. For example, a sporophyte that produces spores that give rise only to male gametophytes may be described as "male", even though the sporophyte itself is asexual, producing only spores. Similarly, flowers produced by the sporophyte may be described as "unisexual" or "bisexual", meaning that they give rise to either one sex of gametophyte or both sexes of gametophyte.

Flowering Plants

Basic Flower Morphology

The flower is the characteristic structure concerned with sexual reproduction in flowering plants (angiosperms). Flowers vary enormously in their construction (morphology). A "complete" flower, like that of *Ranunculus glaberrimus* shown in the figure, has a calyx of outer sepals and a corolla of inner petals. The sepals and petals together form the perianth. Next inwards there are numerous stamens, which produce pollen grains, each containing a microscopic male gametophyte. Stamens may be called the "male" parts of a flower and collectively form the androecium. Finally in the middle there are carpels, which at maturity contain one or more ovules, and within each ovule is a tiny female gametophyte. Carpels may be called the "female" parts of a flower and collectively form the gynoecium.

Flower of *Ranunculus glaberrimus*

Each carpel in *Ranunculus* species is an achene that produces one ovule, which when fertilized becomes a seed. If the carpel contains more than one seed, as in *Eranthis hyemalis* it is called a follicle. Two or more carpels may be fused together to varying degrees and the entire structure, including the fused styles and stigmas may be called a pistil. The lower part of the pistil, where the ovules are produced, is called the ovary. It may be divided into chambers (locules) corresponding to the separate carpels.

Variations

*Alnus serru*lata has unisexual flowers and is monoecious. Shown here: maturing male flower catkins on the right, last year's female catkins on the left.

A "perfect" flower has both stamens and carpels, and may be described as "bisexual" or "hermaphroditic". A "unisexual" flower is one in which either the stamens or the carpels are missing, vestigial or otherwise non-functional. Each flower is either "staminate" (having only functional stamens) and thus "male", or "carpellate" (or "pistillate") (having only functional carpels) and thus "female". If separate staminate and carpellate flowers are always found on the same plant, the species is called monoecious. If separate staminate and carpellate flowers are always found on different plants, the species is called dioecious. A 1995 study found that about 6% of angiosperm species are dioecious, and that 7% of genera contain some dioecious species.

Ilex aquifolium is dioecious: (above) shoot with flowers from male plant; (top right) male flower enlarged, showing stamens with pollen and reduced, sterile stigma; (below) shoot with flowers from female plant; (lower right) female flower enlarged, showing stigma and reduced, sterile stamens (staminodes) with no pollen

Members of the birch family (Betulaceae) are examples of monoecious plants with unisexual flowers. A mature alder tree (*Alnus* species) produces long catkins containing only male flowers, each with four stamens and a minute perianth, and separate stalked groups of female flowers, each without a perianth.

Most hollies (members of the genus *Ilex*) are dioecious. Each plant produces either functionally male flowers or functionally female flowers. In *Ilex aquifolium*, the common European holly, both kinds of flower have four sepals and four white petals; male flowers have four stamens, female flowers usually have four non-functional reduced stamens and a four-celled ovary. Since only female plants are able to set fruit and produce berries, this has consequences for gardeners. *Amborella* represents the first known group of flowering plants to separate from their common ancestor. It too is dioecious; at any one time, each plant produces either flowers with functional stamens but no carpels, or flowers with a few non-functional stamens and a number of fully functional carpels. However, *Amborella* plants may change their "sex" over time. In one study, five cuttings from a male plant produced only male flowers when they first flowered, but at their second flowering three switched to producing female flowers.

In extreme cases, all of the parts present in a complete flower may be missing, so long as at least one carpel or one stamen is present. This situation is reached in the female flowers of duckweeds (*Lemna*), which comprise a single carpel, and in the male flowers of spurges (*Euphorbia*) which comprise a single stamen.

A species such as *Fraxinus excelsior*, the common ash of Europe, demonstrates one possible kind of variation. Ash flowers are wind-pollinated and lack petals and sepals. Structurally, the flowers may be bisexual, consisting of two stamens and an ovary, or may be male (staminate), lacking a functional ovary, or female (carpellate), lacking functional stamens. Different forms may occur on the same tree, or on different trees. The Asteraceae (sunflower family), with close to 22,000 species worldwide, have highly modified inflorescences made up of flowers (florets) collected together into tightly packed heads. Heads may have florets of one sexual morphology – all bisexual, all carpellate or all staminate (when they are called homogamous), or may have mixtures of two or more sexual forms (heterogamous). Thus goatsbeards (*Tragopogon* species) have heads of bisexual florets, like other members of the tribe Cichorieae, whereas marigolds (*Calendula* species) generally have heads with the outer florets bisexual and the inner florets staminate (male).

Like *Amborella*, some plants undergo sex-switching. For example, *Arisaema triphyllum* (Jack-in-the-pulpit) expresses sexual differences at different stages of growth: smaller plants produce all or mostly male flowers; as plants grow larger over the years the male flowers are replaced by more female flowers on the same plant. *Arisaema triphyllum* thus covers a multitude of sexual conditions in its lifetime: nonsexual juvenile plants, young plants that are all male, larger plants with a mix of both male and female flowers, and large plants that have mostly female flowers. Other plant populations have plants that produce more male flowers early in the year and as plants bloom later in the growing season they produce more female flowers.

Terminology

The complexity of the morphology of flowers and its variation within populations has led to a rich terminology.

- Androdioecious: having male flowers on some plants, bisexual ones on others.
- Androecious: having only male flowers (the male of a dioecious population); producing pollen but no seed.
- Androgynous.
- Androgynomonoecious: having male, female, and bisexual flowers on the same plant, also called trimonoecious.
- Andromonoecious: having both bisexual and male flowers on the same plant.
- Bisexual: each flower of each individual has both male and female structures, i.e. it combines both sexes in one structure. Flowers of this kind are called perfect, having both stamens and carpels. Other terms used for this condition are androgynous, hermaphroditic, monoclinous and synoecious.

- Dichogamous: having sexes developing at different times; producing pollen when the stigmas are not receptive, either protandrous or protogynous. This promotes outcrossing by limiting self-pollination. Some dichogamous plants have bisexual flowers, others have unisexual flowers.
- Diclinous.
- Dioecious: having either only male or only female flowers. No individual plant of the population produces both pollen and ovules.
- Gynodioecious: having hermaphrodite flowers and female flowers on separate plants.
- Gynoecious: having only female flowers (the female of a dioecious population); producing seed but not pollen.
- Gynomonoecious: having both bisexual and female flowers on the same plant.
- Hermaphroditic.
- Imperfect: (of flowers) having some parts that are normally present not developed, e.g. lacking stamens.
- Monoclinous.
- Monoecious: In the commoner narrow sense of the term, it refers to plants with unisexual flowers which occur on the same individual. In the broad sense of the term, it also includes plants with bisexual flowers. Individuals bearing separate flowers of both sexes at the same time are called simultaneously or synchronously monoecious. Individuals that bear flowers of one sex at one time are called consecutively monoecious.
- Perfect: (of flowers).
- Polygamodioecious: mostly dioecious, but with either a few flowers of the opposite sex or a few bisexual flowers on the same plant.
- Polygamomonoecious: polygamous. Or, mostly monoecious, but also partly polygamous.
- Polygamous: having male, female, and bisexual flowers on the same plant. Also called polygamomonoecious or trimonoecious. Or, with bisexual and at least one of male and female flowers on the same plant.
- Protandrous: (of dichogamous plants) having male parts of flowers developed before female parts, e.g. having flowers that function first as male and then change to female or producing pollen before the stigmas of the same plant are receptive. (Protoandrous is also used.)

- Protogynous: (of dichogamous plants) having female parts of flowers developed before male parts, e.g. having flowers that function first as female and then change to male or producing pollen after the stigmas of the same plant are receptive.
- Subandroecious: having mostly male flowers, with a few female or bisexual flowers.
- Subdioecious: having some individuals in otherwise dioecious populations with flowers that are not clearly male or female. The population produces normally male or female plants with unisexual flowers, but some plants may have bisexual flowers, some both male and female flowers, and others some combination thereof, such as female and bisexual flowers. The condition is thought to represent a transition between bisexuality and dioecy.
- Subgynoecious: having mostly female flowers, with a few male or bisexual flowers.
- Synoecious.
- Trimonoecious.
- Trioecious.
- Unisexual: having either functionally male or functionally female flowers. This condition is also called diclinous, incomplete or imperfect.

Outcrossing

Outcrossing, cross-fertilization or allogamy, in which offspring are formed by the fusion of the gametes of two different plants, is the most common mode of reproduction among higher plants. About 55% of higher plant species reproduce in this way. An additional 7% are partially cross-fertilizing and partially self-fertilizing (autogamy). About 15% produce gametes but are principally self-fertilizing with significant out-crossing lacking. Only about 8% of higher plant species reproduce exclusively by non-sexual means. These include plants that reproduce vegetatively by runners or bulbils, or which produce seeds without embryo fertilization (apomixis). The selective advantage of outcrossing appears to be the masking of deleterious recessive mutations.

The primary mechanism used by flowering plants to ensure outcrossing involves a genetic mechanism known as self-incompatibility. Various aspects of floral morphology promote allogamy. In plants with bisexual flowers, the anthers and carpels may mature at different times, plants being protandrous (with the anthers maturing first) or protogynous (with the carpels mature first). Monoecious species, with unisexual flowers on the same plant, may produce male and female flowers at different times.

Dioecy, the condition of having unisexual flowers on different plants, necessarily results in outcrossing, and might thus be thought to have evolved for this purpose. However, "dioecy has proven difficult to explain simply as an outbreeding mechanism in plants that lack self-incompatibility". Resource-allocation constraints may be important in the evolution of dioecy, for example, with wind-pollination, separate male flowers arranged in a catkin that vibrates in the wind may provide better pollen dispersal. In climbing plants, rapid upward growth may be essential, and resource allocation to fruit production may be incompatible with rapid growth, thus giving an advantage to delayed production of female flowers. Dioecy has evolved separately in many different lineages, and monoecy in the plant lineage correlates with the evolution of dioecy, suggesting that dioecy can evolve more readily from plants that already produce separate male and female flowers.

Gynoecium

Flower of *Magnolia* × *wieseneri* showing the many pistils making up the gynoecium in the middle of the flower

Hippeastrum flowers showing stamens, style and stigma

Hippeastrum stigmas and style

Moss plants with gynoecia, clusters of archegonia at the apex of each shoot.

Gynoecium is most commonly used as a collective term for the parts of a flower that produce ovules and ultimately develop into the fruit and seeds. The gynoecium is the innermost whorl of (one or more) pistils in a flower and is typically surrounded by the pollen-producing reproductive organs, the stamens, collectively called the androecium.

The gynoecium is often referred to as the "female" portion of the flower, although rather than directly producing female gametes (i.e. egg cells), the gynoecium produces megaspores, each of which develops into a female gametophyte which then produces egg cells.

The term gynoecium is also used by botanists to refer to a cluster of archegonia and any associated modified leaves or stems present on a gametophyte shoot in mosses, liverworts and hornworts. The corresponding terms for the male parts of those plants are clusters of antheridia within the androecium.

Flowers that bear a gynoecium but no stamens are called pistillate or carpellate. Flowers lacking a gynoecium are called staminate.

The gynoecium is often referred to as female because it gives rise to female (egg-producing) gametophytes, however, strictly speaking sporophytes do not have a sex, only gametophytes do.

Gynoecium development and arrangement is important in systematic research and identification of angiosperms, but can be the most challenging of the floral parts to interpret.

Pistils

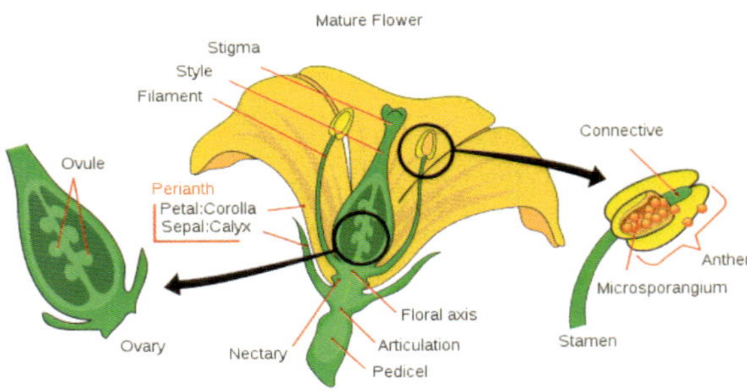

A monocarpous (single carpel) gynoecium in context. The gynoecium (whether composed of a single carpel or multiple "fused" carpels) is typically made up of an ovary, style, and stigma as in the center of the flower.

The gynoecium may consist of one or more separate pistils. A pistil typically consists of an expanded basal portion called the ovary, an elongated section called a style and an apical structure that receives pollen called a stigma.

- The ovary, is the enlarged basal portion which contains placentas, ridges of tissue bearing one or more ovules (integumented megasporangia). The placentas and/or ovule(s) may be born on the gynoecial appendages or less frequently on the floral apex. The chamber in which the ovules develop is called a locule (or sometimes cell).
- The style, is a pillar-like stalk through which pollen tubes grow to reach the ovary. Some flowers such as *Tulipa* do not have a distinct style, and the stigma sits directly on the ovary. The style is a hollow tube in some plants such as lilies, or has transmitting tissue through which the pollen tubes grow.

- The stigma, is usually found at the tip of the style, the portion of the carpel(s) that receives pollen (male gametophytes). It is commonly sticky or feathery to capture pollen.

The word "pistil" comes from Latin *pistillum* meaning pestle. A sterile pistil in a male flower is referred to as a pistillode.

Carpels

The pistils of a flower are considered to be composed of carpels. A carpel is the female reproductive part of the flower, interpreted as modified leaves bearing structures called ovules, inside which the egg cells ultimately form. A pistil may consist of one carpel, with its ovary, style and stigma, or several carpels may be joined together with a single ovary, the whole unit called a pistil. The gynoecium may consist of one or more uni-carpellate (with one carpel) pistils, or of one multi-carpellate pistil. The number of carpels is described by terms such as tricarpellate (three carpels).

Carpels are thought to be phylogenetically derived from ovule-bearing leaves or leaf homologues (megasporophylls), which evolved to form a closed structure containing the ovules. This structure is typically rolled and fused along the margin.

Although many flowers satisfy the above definition of a carpel, there are also flowers that do not have carpels according to this definition because in these flowers the ovule(s), although enclosed, are borne directly on the shoot apex, and only later become enclosed by the carpel. Different remedies have been suggested for this problem. An easy remedy that applies to most cases is to redefine the carpel as an appendage that encloses ovule(s) and may or may not bear them.

Centre of a *Ranunculus repens* (creeping buttercup) showing multiple unfused carpels surrounded by longer stamens

Cross-section through the ovary of *Narcissus* showing multiple connate carpels (a compound pistil) fused along the placental line where the ovules form in each locule

Types

If a gynoecium has a single carpel, it is called monocarpous. If a gynoecium has multiple, distinct (free, unfused) carpels, it is apocarpous. If a gynoecium has multiple carpels "fused" into a single structure, it is syncarpous. A syncarpous gynoecium can sometimes appear very much like a monocarpous gynoecium.

Comparison of gynoecium terminology using *carpel* and *pistil*			
Gynoecium composition	***Carpel* terminology**	***Pistil* terminology**	**Examples**
Single carpel	Monocarpous (unicarpellate) gynoecium	A pistil (simple)	Avocado (*Persea* sp.), most legumes (Fabaceae)
Multiple distinct (unfused) carpels	Apocarpous (choricarpous) gynoecium	Pistils (simple)	Strawberry (*Fragaria* sp.), Buttercup (*Ranunculus* sp.)
Multiple connate ("fused") carpels	Syncarpous gynoecium	A pistil (compound)	Tulip (*Tulipa* sp.), most flowers

The degree of connation ("fusion") in a syncarpous gynoecium can vary. The carpels may be "fused" only at their bases, but retain separate styles and stigmas. The carpels may be "fused" entirely, except for retaining separate stigmas. Sometimes (e.g., Apocynaceae) carpels are fused by their styles or stigmas but possess distinct ovaries. In a syncarpous gynoecium, the "fused" ovaries of the constituent carpels may be referred to collectively as a single compound ovary. It can be a challenge to determine how many carpels fused to form a syncarpous gynoecium. If the styles and stigmas are distinct, they can usually be counted to determine the number of carpels. Within the compound ovary, the carpels may have distinct locules divided by walls called septa. If a syncarpous gynoecium has a single style and stigma and a single locule in the ovary, it may be necessary to examine how the ovules are attached. Each carpel will usually have a distinct line of placentation where the ovules are attached.

Pistil Development

Pistils begin as small primordia on a floral apical meristem, forming later than, and closer to the (floral) apex than sepal, petal and stamen primordia. Morphological and molecular studies of pistil ontogeny reveal that carpels are most likely homologous to leaves.

A carpel has a similar function to a megasporophyll, but typically includes a stigma, and is fused, with ovules enclosed in the enlarged lower portion, the ovary.

In some basal angiosperm lineages, Degeneriaceae and Winteraceae, a carpel begins as a shallow cup where the ovules develop with laminar placentation, on the upper surface of the carpel. The carpel eventually forms a folded, leaf-like structure, not fully sealed at its margins. No style exists, but a broad stigmatic crest along the margin allows pollen tubes access along the surface and between hairs at the margins.

Two kinds of fusion have been distinguished: postgenital fusion that can be observed during the development of flowers, and congenital fusion that cannot be observed i.e., fusions that occurred during phylogeny. But it is very difficult to distinguish fusion and non-fusion processes in the evolution of flowering plants. Some processes that have been considered congenital (phylogenetic) fusions appear to be non-fusion processes such as, for example, the de novo formation of intercalary growth in a ring zone at or below the base of primordia. Therefore, "it is now increasingly acknowledged that the term 'fusion,' as applied to phylogeny (as in 'congenital fusion') is ill-advised."

Gynoecium Position

Flowers and fruit (capsules) of the ground orchid, *Spathoglottis plicata*, illustrating an **inferior** ovary.

Basal angiosperm groups tend to have carpels arranged spirally around a conical or dome-shaped receptacle. In later lineages, carpels tend to be in whorls.

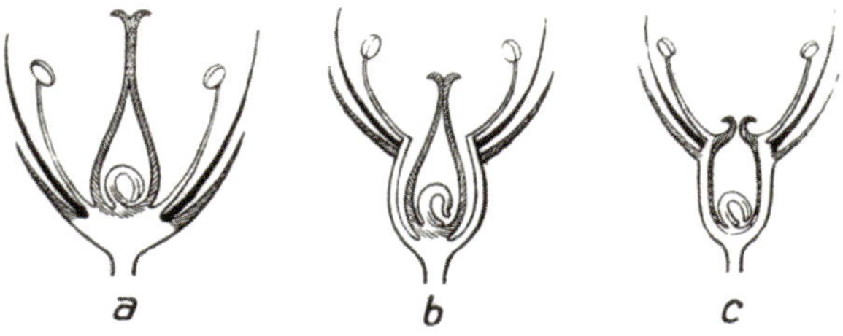

Illustration showing longitudinal sections through hypogynous (a), perigynous (b), and epigynous (c) flowers

The relationship of the other flower parts to the gynoecium can be an important systematic and taxonomic character. In some flowers, the stamens, petals, and sepals are often said to be "fused" into a "floral tube" or hypanthium. However, as Leins & Erbar (2010) pointed out, "the classical view that the wall of the inferior ovary results from the "congenital" fusion of dorsal carpel flanks and the floral axis does not correspond to the ontogenetic processes that can actually be observed. All that can be seen is an intercalary growth in a broad circular zone that changes the shape of the floral axis (receptacle)." And what happened during evolution is not a phylogenetic fusion but the formation of a unitary intercalary meristem. Evolutionary developmental biology investigates such developmental processes that arise or change during evolution.

If the hypanthium is absent, the flower is *hypogynous*, and the stamens, petals, and sepals are all attached to the receptacle below the gynoecium. Hypogynous flowers are often referred to as having a *superior ovary*. This is the typical arrangement in most flowers.

If the hypanthium is present up to the base of the style(s), the flower is *epigynous*. In an epigynous flower, the stamens, petals, and sepals are attached to the hypanthium at the top of the ovary or, occasionally, the hypanthium may extend beyond the top of the ovary. Epigynous flowers are often referred to as having an *inferior ovary*. Plant families with epigynous flowers include orchids, asters, and evening primroses.

Between these two extremes are *perigynous* flowers, in which a hypanthium is present, but is either free from the gynoecium (in which case it may appear to be a cup or tube surrounding the gynoecium) or connected partly to the gynoecium (with the stamens, petals, and sepals attached to the hypanthium part of the way up the ovary). Perigynous flowers are often referred to as having a *half-inferior ovary* (or, sometimes, *partially inferior* or *half-superior*). This arrangement is particularly frequent in the rose family and saxifrages.

Occasionally, the gynoecium is born on a stalk, called the gynophore, as in *Isomeris arborea*.

Placentation

Within the ovary, each ovule is born by a placenta or arises as a continuation of the floral apex. The placentas often occur in distinct lines called lines of placentation. In monocarpous or apocarpous gynoecia, there is typically a single line of placentation in each ovary. In syncarpous gynoecia, the lines of placentation can be regularly spaced along the wall of the ovary (parietal placentation), or near the center of the ovary. In the latter case, separate terms are used depending on whether or not the ovary is divided into separate locules. If the ovary is divided, with the ovules born on a line of placentation at the inner angle of each locule, this is axile placentation. An ovary with free central placentation, on the other hand, consists of a single compartment without septae and the ovules are attached to a central column that arises directly from the floral apex (axis). In some cases a single ovule is attached to the bottom or top of the locule (basal or apical placentation, respectively).

The Ovule

In flowering plants, the *ovule* is a complex structure born inside ovaries. The ovule initially consists of a stalked, integumented *megasporangium* (also called the *nucellus*). Typically, one cell in the megasporangium undergoes meiosis resulting in one to four megaspores. These develop into a megagametophyte (often called the embryo sac) within the ovule. The megagametophyte typically develops a small number of cells, including two special cells, an egg cell and a binucleate central cell, which are the gametes involved in double fertilization. The central cell, once fertilized by a sperm cell from the pollen becomes the first cell of the endosperm, and the egg cell once fertilized become the zygote that develops into the embryo. The gap in the integuments through which the pollen tube enters to deliver sperm to the egg is called the micropyle. The stalk attaching the ovule to the placenta is called the funiculus.

Role of the Stigma and Style

Stigmas and style of *Cannabis sativa* held in a pair of forceps

Stigma of a *Crocus* flower

Stigmas can vary from long and slender to globe-shaped to feathery. The stigma is the receptive tip of the carpel(s), which receives pollen at pollination and on which the pollen grain germinates. The stigma is adapted to catch and trap pollen, either by combining pollen of visiting insects or by various hairs, flaps, or sculpturings.

The style and stigma of the flower are involved in most types of self incompatibility reactions. Self-incompatibility, if present, prevents fertilization by pollen from the same plant or from genetically similar plants, and ensures outcrossing.

Stigma (Botany)

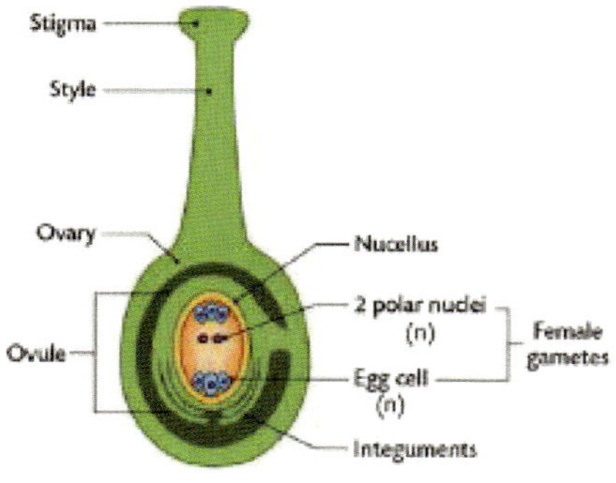

Diagram of stigma

The stigma (plural: stigmata) is the receptive tip of a carpel, or of several fused carpels, in the gynoecium of a flower.

Description

The stigma, together with the style and ovary comprises the pistil, which in turn is part of the gynoecium or female reproductive organ of a plant. The stigma forms the distal portion of the style or stylodia. The stigma is composed of stigmatic papillae, the cells which are receptive to pollen. These may be restricted to the apex of the style or, especially in wind pollinated species, cover a wide surface.

Stigma of a *Tulipa* species, with pollen

The stigma receives pollen and it is on the stigma that the pollen grain germinates. Often sticky, the stigma is adapted in various ways to catch and trap pollen with various hairs, flaps, or sculpturings. The pollen may be captured from the air (wind-borne pollen, anemophily), from visiting insects or other animals (biotic pollination), or in rare cases from surrounding water (hydrophily). Stigma can vary from long and slender to globe shaped to feathery.

Pollen is typically highly desiccated when it leaves an anther. Stigma have been shown to assist in the rehydration of pollen and in promoting germination of the pollen tube. Stigma also ensure proper adhesion of the correct species of pollen. Stigma can play an active role in pollen discrimination and some self-incompatibility reactions, that reject pollen from the same or genetically similar plants, involve interaction between the stigma and the surface of the pollen grain.

Shape

The stigma is often split into lobes, e.g. trifid (three lobed), and may resemble the head of a pin (capitate), or come to a point (punctiform). The shape of the stigma may vary considerably:

Style

Structure

The style is a narrow upward extension of the ovary, connecting it to the stigmatic papillae. It may be absent in some plants in the case the stigma is referred to as sessile. Styles are generally tube-like—either long or short. The style can be open (containing few or no cells in the central portion) with a central canal which may be filled with mucilage. Alternatively the style may be closed (densely packed with cells throughout). Most syncarpous monocots and some eudicots have open styles, while many syncarpous eudicots and grasses have closed (solid) styles containing specialised secretory transmitting tissue tissue, linking the stigma to the centre of the ovary. This forms a nutrient rich tract for pollen tube growth.

Corn stigma called "silk"

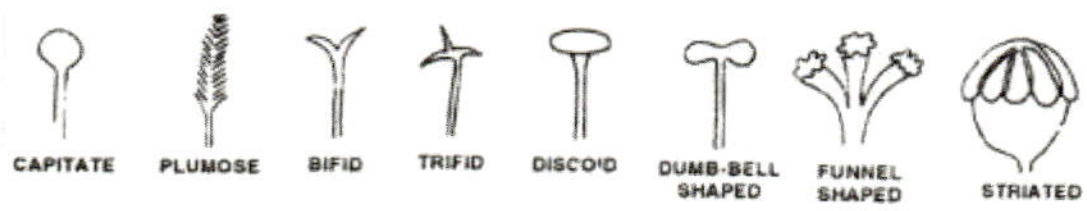

Types of Stigma

Where there are more than one carpels to the pistil, each may have a separate style-like stylodium, or share a common style. In Irises and others in the Iridaceae family, the style divides into three petal-like (petaloid) *style branches* (sometimes also referred to as 'stylodia'), almost to the base of the style and is called tribrachiate. These are flaps of tissue, running from the perianth tube above the sepal. The stigma is a rim or edge on the underside of the branch, near the end lobes. Style branches also appear on *Dietes*, *Pardanthopsis* and most species of *Moraea*.

In *Crocuses*, there are three divided style branches, creating a tube. *Hesperantha* has a spreading style branch. Alternatively the style may be lobed rather than branched. *Gladiolus* has a bi-lobed style branch (bilobate). *Freesia*, *Lapeirousia*, *Romulea*, *Savannosiphon* and *Watsonia* have bifuracated (two branched) and recurved style branches.

Style Morphology

Iris versicolor showing three structures with two overlapping lips, an upper petaloid style branch and a lower tepal, enclosing a stamen

Iris missouriensis showing the pale blue style branch above the drooping petal

The feathery stigma of *Crocus speciosus* has branches corresponding to three carpels

Attachment to the Ovary

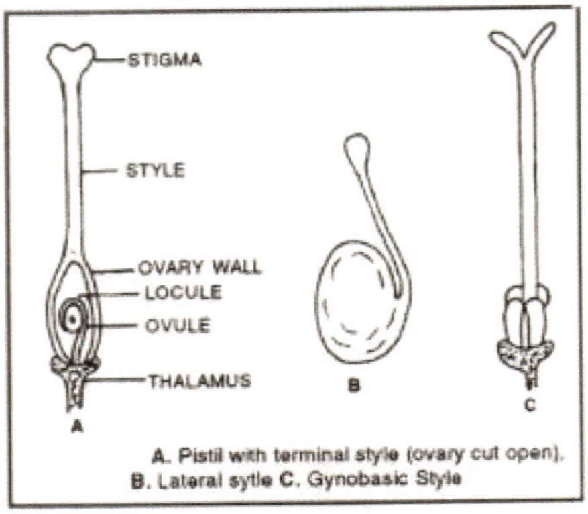

A. Pistil with terminal style (ovary cut open). B. Lateral sytle C. Gynobasic Style

May be terminal (apical), subapical, lateral, gynobasic, or subgynobasic. Terminal (apical) style position refers to attachment at the apex of the ovary and is the commonest pattern. In the subapical pattern the style arises to the side slightly below the apex. a lateral style arises from the side of the ovary and is found in Rosaceae. The gynobasic style arises from the base of the ovary, or between the ovary lobes and is characteristic of Boraginaceae. Subgynobasic styles characterise *Allium*.

Pollination

Pollen tubes grow the length of the style to reach the ovules, and in some cases self-incompatibility reactions in the style prevent full growth of the pollen tubes. In some species, including *Gasteria* at least, the pollen tube is directed to the micropyle of the ovule by the style.

Ovary (Botany)

Longitudinal section of female flower of squash showing pistil (=ovary+style+stigma), ovules, and petals. The petals and sepals are above the ovary; such a flower is said to have an inferior ovary, or the flower is said to be epigynous.

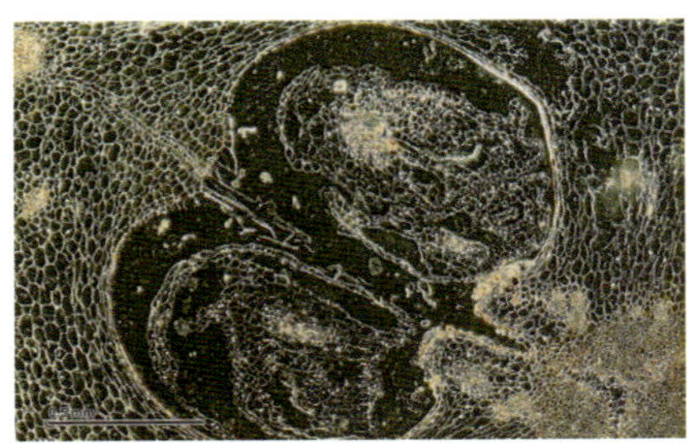
Cross section of *Tulip* ovary

In the flowering plants, an ovary is a part of the female reproductive organ of the flower or gynoecium. Specifically, it is the part of the pistil which holds the ovule(s) and is located above or below or at the point of connection with the base of the petals and sepals. The pistil may be made up of one carpel or of several fused carpels (e.g. dicarpel or tricarpel), and therefore the ovary can contain part of one carpel or parts of several fused carpels. Above the ovary is the style and the stigma, which is where the pollen lands and germinates to grow down through the style to the ovary, and, for each individual pollen grain, to fertilize one individual ovule. Some wind pollinated flowers have much reduced and modified ovaries.

Fruits

A fruit is the ripened ovary or ovaries—together with seeds—from one or more flowers. The fruits of a plant are responsible for dispersing the seeds that contain the embryo and protecting the seeds as well. In many species, the fruit incorporates some surrounding tissues, or is dispersed with some non-fruit tissues.

Parts of the Ovary

Locules are chambers within the ovary of the flower and fruits. The locules contain the ovules (seeds), and may or may not be filled with fruit flesh. Depending on the number of locules in the ovary, fruits can be classified as uni-locular (unilocular), bi-locular, tri-locular or multi-locular. Some plants have septa between the carpels; the number of locules present in a gynoecium may be equal to or less than the number of carpels, depending on whether septa are present.

The ovules are attached to parts of the interior ovary walls called the placentae. Placental areas occur in various positions, corresponding to various parts of the carpels that make up the ovary. An obturator is present in the ovary of some plants, near the micropyle of each ovule. It is an outgrowth of the placenta, important in nourishing and guiding pollen tubes to the micropyle.

The ovary of some types of fruit is dehiscent; the ovary wall splits into sections called valves. There is no standard correspondence between the valves and the position of the septa; the valves may separate by splitting the septa (septicidal dehiscence) or by spitting between them (loculicidal dehiscence), or the ovary may open in other ways such as through pores or because a cap falls off.

The syncarpous ovary of this melon is made up of four carpels, and has one locule.

In this *Peganum harmala*, the ovary of a fruit has split into valves.

The valves of *Lunaria* fruit fall to reveal a septum that was between the two carpels of the ovary.

The seeds in a tomato fruit grow from placental areas at the interior of the ovary. (This is axile placentation in a bi-locular fruit.)

The placentae in *Lunaria* are along the margins of the fruit, where two carpels fuse. (This is parietal placentation in a bi-locular fruit.)

Classification Based on Position

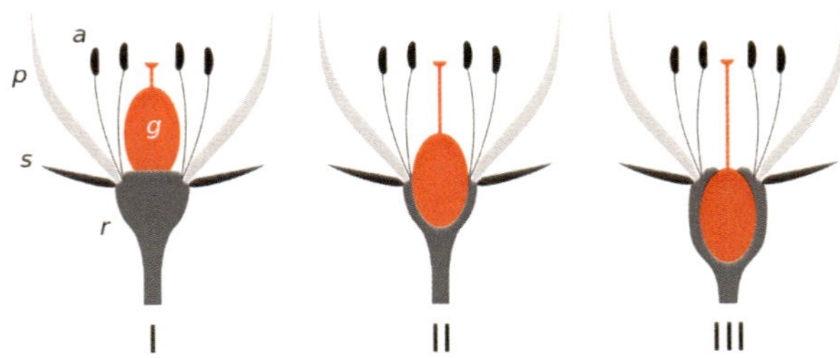

Ovary insertion: I superior II half-inferior III inferior. *a* androecium *g* gynoecium *p* petals *s* sepals *r* receptacle. The insertion point is where *a*, *p*, and *s* converge.

The terminology of the positions of ovaries is determined by the insertion point, where the other floral parts (perianth and androecium) come together and attach to the surface of the ovary. If the ovary is situated above the insertion point, it is superior; if below, inferior.

Superior Ovary

A superior ovary is an ovary attached to the receptacle above the attachment of other floral parts. A superior ovary is found in types of fleshy fruits such as true berries, drupes, etc. A flower with this arrangement is described as hypogynous. Examples of this ovary type include the legumes (beans and peas and their relatives).

Half-inferior Ovary

A half-inferior ovary (also known as "half-superior", "subinferior," or "partially inferior,") is embedded or surrounded by the receptacle. This occurs in flowers of the Lythraceae family, which includes the Crape Myrtles. Such flowers are termed perigynous or half-epigynous. In some classifications, half-inferior ovaries are not recognized and are instead grouped with either the superior or inferior ovaries.

More specifically, a half-inferior ovary has nearly equal portions of ovary above and below the insertion point. Other varying degrees of inferiority can be described by other fractions. For instance, a "one-fifth inferior ovary" has approximately one fifth of its length under the insertion point. Likewise, only one quarter portion of a "three-quarters inferior ovary" is above the insertion.

Inferior Ovary

An inferior ovary lies below the attachment of other floral parts. A pome is a type of fleshy fruit that is often cited as an example, but close inspection of some pomes (such as *Pyracantha*) will show that it is really a half-inferior ovary. Flowers with inferior ovaries are termed epigynous. Some examples of flowers with an inferior ovary are orchids (inferior capsule), *Fuchsia* (inferior berry), banana (inferior berry), Asteraceae (inferior achene-like fruit, called a cypsela) and the pepo of the squash, melon and gourd (Cucurbitaceae) family.

Ovule

Location of ovules inside a *Helleborus foetidus* flower

In seed plants, the ovule ("small egg") is the structure that gives rise to and contains the female reproductive cells. It consists of three parts: The integument(s), forming its outer layer(s), the nucellus (or remnant of the megasporangium), and the female gametophyte (formed from a haploid megaspore) in its center. The female gametophyte — specifically termed a *megagametophyte*— is also called the embryo sac in angiosperms. The megagametophyte produces an egg cell (or several in some groups) for the purpose of fertilization.

Location within the Plant

In flowering plants, the ovule is located inside the portion of the flower called the gynoecium. The ovary of the gynoecium produces one or more ovules and ultimately becomes the fruit wall. Ovules are attached to the placenta in the ovary through a stalk-like structure known as a funiculus (plural, funiculi). Different patterns of ovule attachment, or placentation, can be found among plant species, these include:

- Apical placentation: The placenta is at the apex (top) of the ovary. Simple or compound ovary

- Axile placentation: The ovary is divided into radial segments, with placentas in separate locules. Ventral sutures of carpels meet at the centre of the ovary. Placentae are along fused margins of carpels. Two or more carpels. (e.g. *Hibiscus*, *Citrus*, *Solanum*)
- Basal placentation: The placenta is at the base (bottom) of the ovary on a protrusion of the thalamus (receptacle). Simple or compound carpel, unilocular ovary. (e.g. *Sonchus*, *Helianthus*, Compositae)
- Free-central placentation: Derived from axile as partitions are absorbed, leaving ovules at the central axis. Compound unilocular ovary. (e.g. *Stellaria*, *Dianthus*)
- Marginal placentation: Simplest type. There is only one elongated placenta on one side of the ovary, as ovules are attached at the fusion line of the carpel's margins . This is conspicuous in legumes. Simple carpel, unilocular ovary. (e.g. *Pisum*)
- Parietal placentation: Placentae on inner ovary wall within a non-sectioned ovary, corresponding to fused carpel margins. Two or more carpels, unilocular ovary. (e.g. *Brassica*)
- Superficial: Similar to axile, but placentae are on inner surfaces of multilocular ovary (e.g. *Nymphaea*)

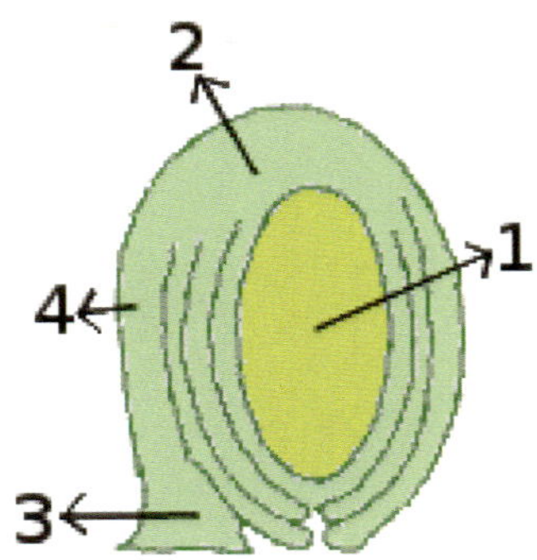

Ovule structure (anatropous) 1: nucleus 2: chalaza 3: funiculus 4: raphe

Ovule orientation may be anatropous, such that when inverted the micropyle faces the placenta (this is the most common ovule orientation in flowering plants), amphitropous, campylotropous, or orthotropous.

In gymnosperms such as conifers, ovules are borne on the surface of an ovuliferous (ovule-bearing) scale, usually within an ovulate cone (also called megastrobilus). In some extinct plants (e.g. Pteridosperms), megasporangia and perhaps ovules were borne on the surface of leaves. In other extinct taxa, a cupule (a modified leaf or part of a leaf) surrounds the ovule (e.g. *Caytonia* or *Glossopteris*).

Ovule Parts and Development

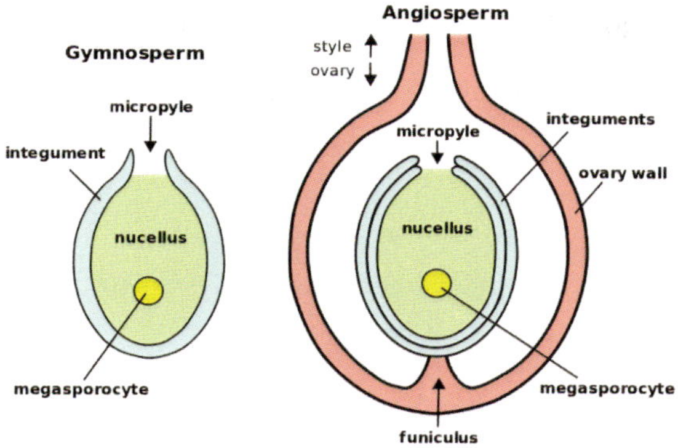

Plant ovules: Gymnosperm ovule on left, angiosperm ovule (inside ovary) on right

Models of different ovules, Botanical Museum Greifswald

The ovule appears to be a megasporangium with integuments surrounding it. Ovules are initially composed of diploid maternal tissue, which includes a megasporocyte (a cell that will undergo meiosis to produce megaspores). Megaspores remain inside the ovule and divide by mitosis to produce the haploid female gametophyte or megagametophyte, which also remains inside the ovule. The remnants of the megasporangium tissue (the nucellus) surround the megagametophyte. Megagametophytes produce archegonia (lost in some groups such as flowering plants), which produce egg cells. After fertilization, the ovule contains a diploid zygote and then, after cell division begins, an embryo of the next sporophyte generation. In flowering plants, a second sperm nucleus fuses with other nuclei in the megagametophyte forming a typically polyploid (often triploid) endosperm tissue, which serves as nourishment for the young sporophyte.

Integuments, Micropyle and Chalaza

An integument is a protective cell layer surrounding the ovule. Gymnosperms typically have one integument (unitegmic) while angiosperms typically have two (bitegmic). The evolutionary origin of the inner integument (which is integral to the formation of ovules from megasporangia) has been proposed to be by enclosure of a megasporangium by sterile branches (telomes). *Elkinsia*, a preovulate taxon, has a lobed structure fused to the lower third of the megasporangium, with the lobes extending upwards in a ring around the megasporangium. This might, through fusion between lobes and between the structure and the megasporangium, have produced an integument.

The origin of the second or outer integument has been an area of active contention for some time. The cupules of some extinct taxa have been suggested as the origin of the outer integument. A few angiosperms produce vascular tissue in the outer integument, the orientation of which suggests that the outer surface is morphologically abaxial. This suggests that cupules of the kind produced by the Caytoniales or Glossopteridales may have evolved into the outer integument of angiosperms.

The integuments develop into the seed coat when the ovule matures after fertilization.

The integuments do not enclose the nucellus completely but retain an opening at the apex referred to as the micropyle. The micropyle opening allows the pollen (a male gametophyte) to enter the ovule for fertilization. In gymnosperms (e.g., conifers), the pollen is drawn into the ovule on a drop of fluid that exudes out of the micropyle, the so-called pollination drop mechanism. Subsequently, the micropyle closes. In angiosperms, only a pollen tube enters the micropyle. During germination, the seedling's radicle emerges through the micropyle.

Located opposite from the micropyle is the chalaza where the nucellus is joined to the integuments. Nutrients from the plant travel through the phloem of the vascular system to the funiculus and outer integument and from there apoplastically and symplastically through the chalaza to the nucellus inside the ovule. In chalazogamous plants, the pollen tubes enter the ovule through the chalaza instead of the micropyle opening.

Nucellus, Megaspore and Perisperm

The nucellus (plural: nucelli) is part of the inner structure of the ovule, forming a layer of diploid (sporophytic) cells immediately inside the integuments. It is structurally and functionally equivalent to the megasporangium. In immature ovules, the nucellus contains a megasporocyte (megaspore mother cell), which undergoes sporogenesis via meiosis. In the megasporocyte of *Arabidopsis thaliana*, meiosis depends on the expression of genes that facilitate DNA repair and homologous recombination.

In gymnosperms, three of the four haploid spores produced in meiosis typically degenerate, leaving one surviving megaspore inside the nucellus. Among angiosperms, however, a wide range of variation exists in what happens next. The number (and position) of surviving megaspores, the total number of cell divisions, whether nuclear fusions occur, and the final number, position and ploidy of the cells or nuclei all vary. A common pattern of embryo sac development (the *Polygonum* type maturation pattern) includes a single functional megaspore followed by three rounds of mitosis. In some cases, however, two megaspores survive (for example, in *Allium* and *Endymion*). In some cases all four megaspores survive, for example in the *Fritillaria* type of development (illustrated by *Lilium* in the figure) there is no separation of the megaspores following meiosis, then the nuclei fuse to form a triploid nucleus and a haploid nucleus. The subsequent arrangement of cells is similar to the *Polygonum* pattern, but the ploidy of the nuclei is different.

After fertilization, the nucellus may develop into the perisperm that feeds the embryo. In some plants, the diploid tissue of the nucellus can give rise to the embryo within the seed through a mechanism of asexual reproduction called nucellar embryony.

Megagametophyte

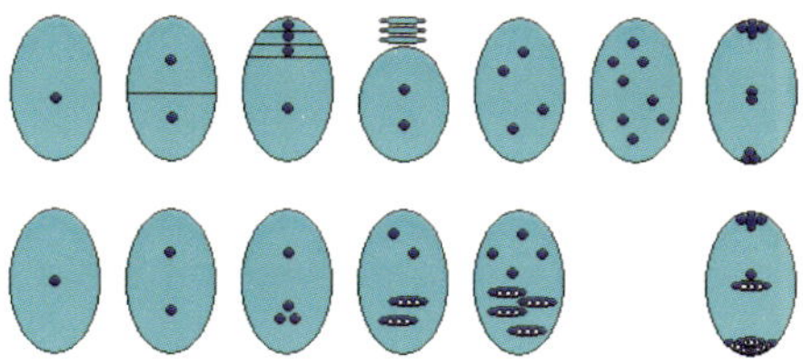

Megagametophyte formation of the genera *Polygonum* and *Lilium*. Triploid nuclei are shown as ellipses with three white dots. The first three columns show the meiosis of the megaspore, followed by 1-2 mitoses.

The haploid megaspore inside the nucellus gives rise to the female gametophyte, called the megagametophyte.

In gymnosperms, the megagametophyte consists of around 2000 nuclei and forms archegonia, which produce egg cells for fertilization.

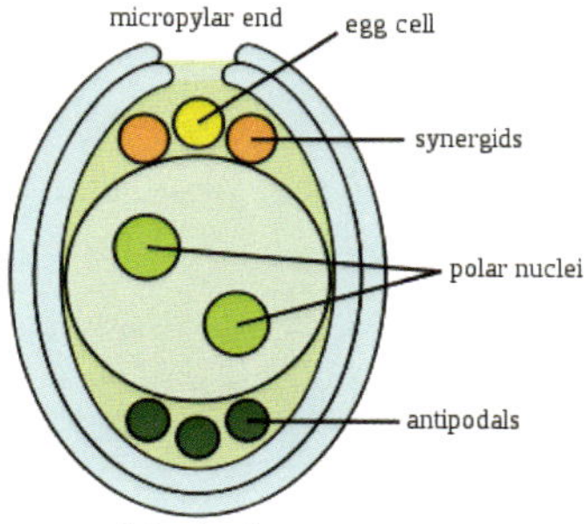

Ovule with megagametophyte: egg cell (yellow), synergids (orange), central cell with two polar nuclei (bright green), and antipodals (dark green)

In flowering plants, the megagametophyte (also referred to as the embryo sac) is much smaller and typically consists of only seven cells and eight nuclei. This type of megagametophyte develops from the megaspore through three rounds of mitotic divisions. The cell closest to the micropyle opening of the integuments differentiates into the egg cell, with two synergid cells by its side that are involved in the production of signals that guide the pollen tube. Three antipodal cells form on the opposite (chalazal) end of the ovule and later degenerate. The large central cell of the embryo sac contains two polar nuclei.

Zygote, Embryo and Endosperm

The pollen tube releases two sperm nuclei into the ovule. In gymnosperms, fertilization occurs within the archegonia produced by the female gametophyte. While it is possible that several egg cells are present and fertilized, typically only one zygote will develop into a mature embryo as the resources within the seed are limited.

In flowering plants, one sperm nucleus fuses with the egg cell to produce a zygote, the other fuses with the two polar nuclei of the central cell to give rise to the polyploid (typically triploid) endosperm. This double fertilization is unique to flowering plants, although in some other groups the second sperm cell does fuse with another cell in the megagametophyte to produce a second embryo. The plant stores nutrients such as starch, proteins, and oils in the endosperm as a food source for the developing embryo and seedling, serving a similar function to the yolk of animal eggs. The endosperm is also called the albumen of the seed.

Embryos may be described by a number of terms including Linear (embryos have axile placentation and are longer than broad), or rudimentary (embryos are basal in which the embryo is tiny in relation to the endosperm).

Types of Gametophytes

Megagametophytes of flowering plants may be described according to the number of megaspores developing, as either monosporic, bisporic, or tetrasporic.

Stamen

Stamens of a *Hippeastrum* with white filaments and prominent anthers carrying pollen

The stamen (plural *stamina* or *stamens*) is the pollen-producing reproductive organ of a flower. Collectively the stamens form the androecium.

Morphology and Terminology

A stamen typically consists of a stalk called the filament and an anther which contains *microsporangia*. Most commonly anthers are two-lobed and are attached to the filament either at the base or in the middle area of the anther. The sterile tissue between the lobes is called the connective. A pollen grain develops from a microspore in the microsporangium and contains the male gametophyte.

The stamens in a flower are collectively called the *androecium*. The androecium can consist of as few as one-half stamen (i.e. a single locule) as in *Canna* species or as many as 3,482 stamens which have been counted in *Carnegiea gigantea*. The androecium in various species of plants forms a great variety of patterns, some of them highly complex. It surrounds the gynoecium and is surrounded by the perianth. A few members of the family Triuridaceae, particularly *Lacandonia schismatica*, are exceptional in that their gynoecia surround their androecia.

Variation in Morphology

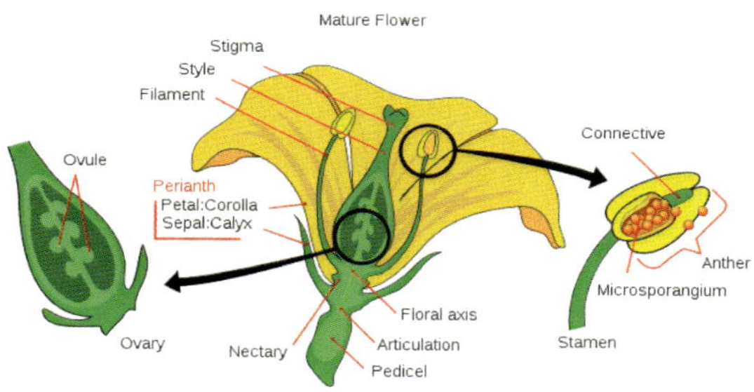

Stamens, with distal anther attached to the filament stalk, in context of floral anatomy

Depending on the species of plant, some or all of the stamens in a flower may be attached to the petals or to the floral axis. They also may be free-standing or fused to one another in many different ways, including fusion of some but not all stamens. The filaments may be fused and the anthers free, or the filaments free and the anthers fused. Rather than there being two locules, one locule of a stamen may fail to develop, or alternatively the two locules may merge late in development to give a single locule. Extreme cases of stamen fusion occur in some species of *Cyclanthera* in the family Cucurbitaceae and in section *Cyclanthera* of genus *Phyllanthus* (family Euphorbiaceae) where the stamens form a ring around the gynoecium, with a single locule.

Cross section of a *Lilium* stamen, with four locules surrounded by the tapetum

Pollen Production

A typical anther contains four microsporangia. The *microsporangia* form sacs or pockets (*locules*) in the anther (anther sacs or pollen sacs). The two separate locules on each side of an anther may fuse into a single locule. Each microsporangium is lined with a nutritive tissue layer called the *tapetum* and initially contains diploid pollen mother cells. These undergo meiosis to form haploid spores. The spores may remain attached to each other in a tetrad or separate after meiosis. Each microspore then divides mitotically to form an immature microgametophyte called a pollen grain.

The pollen is eventually released when the anther forms openings (dehisces). These may consist of longitudinal slits, pores, as in the heath family (Ericaceae), or by valves, as in the barberry family (Berberidaceae). In some plants, notably members of Orchidaceae and Asclepiadoideae, the pollen remains in masses called pollinia, which are adapted to attach to particular pollinating agents such as birds or insects. More commonly, mature pollen grains separate and are dispensed by wind or water, pollinating insects, birds or other pollination vectors.

Pollen of angiosperms must be transported to the stigma, the receptive surface of the *carpel*, of a compatible flower, for successful pollination to occur. After arriving, the pollen grain (an immature microgametophyte) typically completes its development. It may grow a pollen tube and undergoing mitosis to produce two sperm nuclei.

Sexual Reproduction in Plants

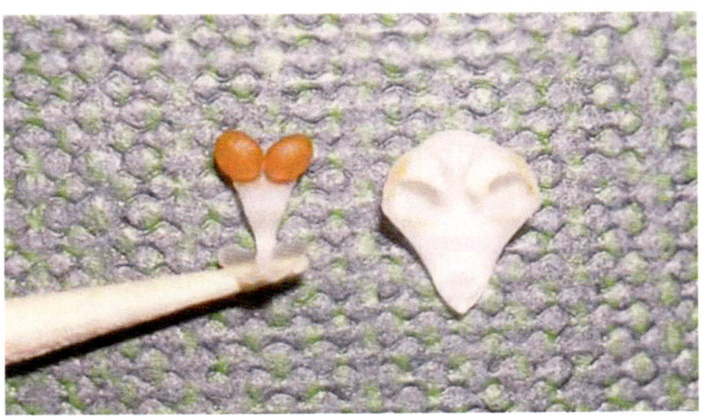

Stamen with pollinia and its anther cap. *Phalaenopsis* orchid.

In the typical flower (that is, in the majority of flowering plant species) each flower has both carpels and stamens. In some species, however, the flowers are unisexual with only carpels or stamens. (monoecious = both types of flowers found on the same plant; dioecious = the two types of flower found only on different plants). A flower with only stamens is called androecious. A flower with only carpels is called gynoecious.

A flower having only functional stamens and lacking functional carpels is called a staminate flower, or (inaccurately) male. A plant with only functional carpels is called pistillate, or (inaccurately) female.

An abortive or rudimentary stamen is called a staminodium or staminode, such as in *Scrophularia nodosa*.

The carpels and stamens of orchids are fused into a column. The top part of the column is formed by the anther, which is covered by an anther cap.

Descriptive Terms

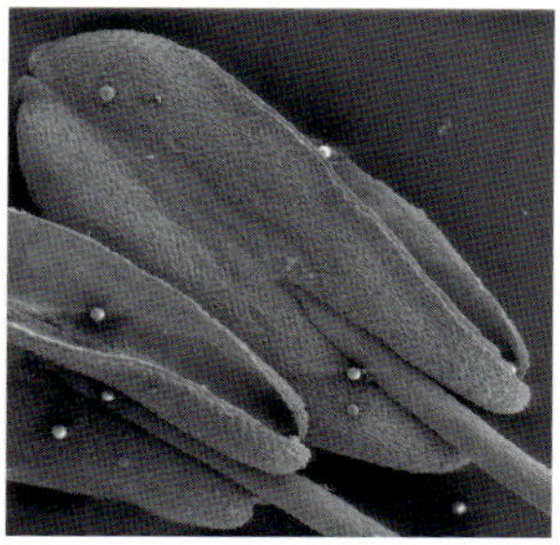

Scanning electron microscope image of *Pentas lanceolata* anthers, with pollen grains on surface

Lily stamens with prominent red anthers and white filaments

- A column formed from the fusion of multiple filaments is known as an *androphore*

The anther can be attached to the filament's connective in two ways:

- basifixed: attached at its base to the filament
    - pseudobasifixed: a somewhat misnomer configuration where connective tissue extends in a tube around the filament
- dorsifixed: attached at its center to the filament, usually versatile (able to move)

Stamens can be connate (fused or joined in the same whorl):

- extrorse: anther dehiscence directed away from the centre of the flower. Cf. introrse, directed inwards, and latrorse towards the side
- monadelphous: fused into a single, compound structure
- declinate: curving downwards, then up at the tip (also – declinate-descending)
- diadelphous: joined partially into two androecial structures
- pentadelphous: joined partially into five androecial structures
- synandrous: only the anthers are connate (such as in the Asteraceae). The fused stamens are referred to as a synandrium

Stamens can also be adnate (fused or joined from more than one whorl):

- epipetalous: adnate to the corolla
- epiphyllous: adnate to undifferentiated tepals (as in many Liliaceae)

They can have different lengths from each other:

- didymous: two equal pairs
- didynamous: occurring in two pairs, a long pair and a shorter pair
- tetradynamous: occurring as a set of six stamens with four long and two shorter ones

or respective to the rest of the flower (perianth):

- exserted: extending beyond the corolla
- included: not extending beyond the corolla

They may be arranged in one of two different patterns:

- spiral; or
- whorled: one or more discrete whorls (series)

They may be arranged, with respect to the petals:

- diplostemonous: in two whorls, the outer alternating with the petals, while the inner is opposite the petals
- obdiplostemonous: in two whorls, the outer opposite the petals

Pollen

Tulip anther with many grains of pollen

Closeup image of a cactus flower and its stamens

Pollen is a fine to coarse powdery substance comprising pollen grains which are male microgametophytes of seed plants, which produce male gametes (sperm cells). Pollen grains have a hard coat made of sporopollenin that protects the gametophytes during the process of their movement from the stamens to the pistil of flowering plants, or from the male cone to the female cone of coniferous plants. If pollen lands on a compatible pistil or female cone, it germinates, producing a pollen tube that transfers the sperm to the ovule containing the female gametophyte. Individual pollen grains are small enough to require magnification. The study of pollen is called palynology and is highly useful in paleoecology, paleontology, archaeology, and forensics.

Scanning electron microscope image (500x magnification) of pollen grains from a variety of common plants: sunflower (*Helianthus annuus*), morning glory (*Ipomoea purpurea*), prairie hollyhock (*Sidalcea malviflora*), oriental lily (*Lilium auratum*), evening primrose (*Oenothera fruticosa*), and castor bean (*Ricinus communis*).

Pollen in plants is used for transferring haploid male genetic material from the anther of a single flower to the stigma of another in cross-pollination. In a case of self-pollination, this process takes place from the anther of a flower to the stigma of the same flower.

The Structure and Formation of Pollen

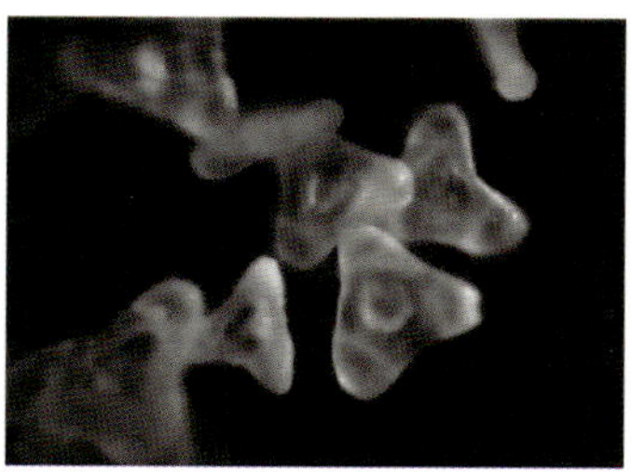

Triporate pollen of *Oenothera speciosa*

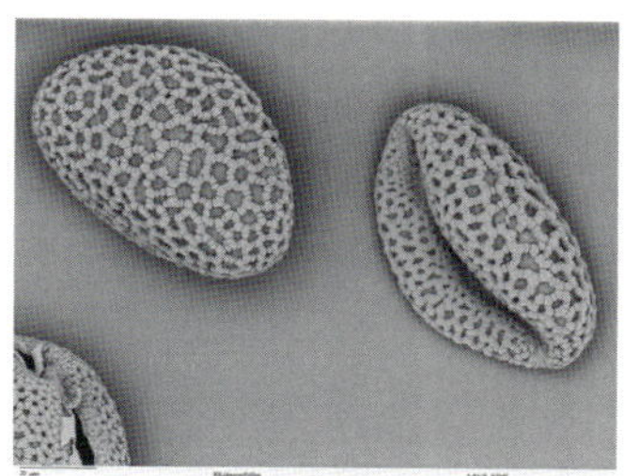

Pollen of *Lilium auratum* showing single sulcus (monosulcate)

Arabis pollen has three colpi and prominent surface structure.

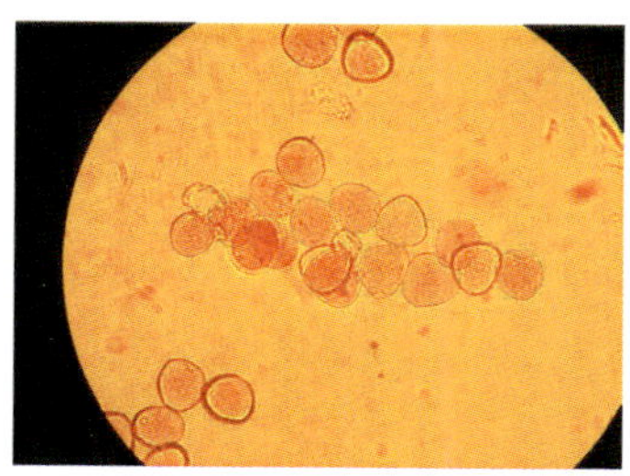

Apple pollen under microscopy

Pollen itself is not the male gamete. Each pollen grain contains vegetative (non-reproductive) cells (only a single cell in most flowering plants but several in other seed plants) and a generative (reproductive) cell. In flowering plants the vegetative tube cell produces the pollen tube, and the generative cell divides to form the two sperm cells.

Formation

Pollen is produced in the microsporangia in the male cone of a conifer or other gymnosperm or in the anthers of an angiosperm flower. Pollen grains come in a wide variety of shapes, sizes, and surface markings characteristic of the species. Pollen grains of pines, firs, and spruces are winged. The smallest pollen grain, that of the forget-me-not (*Myosotis* spp.), is around 6 µm (0.006 mm) in diameter. Wind-borne pollen grains can be as large as about 90–100 µm.

In angiosperms, during flower development the anther is composed of a mass of cells that appear undifferentiated, except for a partially differentiated dermis. As the flower develops, four groups of sporogenous cells form within the anther. The fertile sporogenous cells are surrounded by layers of sterile cells that grow into the wall of the pollen sac. Some of the cells grow into nutritive cells that supply nutrition for the microspores that form by meiotic division from the sporogenous cells.

In a process called microsporogenesis, four haploid microspores are produced from each diploid sporogenous cell (microsporocyte, pollen mother cell or meiocyte), after meiotic division. After the formation of the four microspores, which are contained by callose walls, the development of the pollen grain walls begins. The callose wall is broken down by an enzyme called callase and the freed pollen grains grow in size and develop their characteristic shape and form a resistant outer wall called the exine and an inner wall called the intine. The exine is what is preserved in the fossil record. Two basic types of microsporogenesis are recognised, simultaneous and successive. In simultaneous microsporogenesis meiotic steps I and II are completed prior to cytokinesis, whereas in successive microsporogenesis cytokinesis follows. While there may be a continuum with intermediate forms, the type of microsporogenesis has systematic significance. The predominant form amongst the monocots is successive, but there are important exceptions.

During microgametogenesis, the unicellular microspores undergo mitosis and develop into mature microgametophytes containing the gametes. In some flowering plants, germination of the pollen grain may begin even before it leaves the microsporangium, with the generative cell forming the two sperm cells.

Structure

Except in the case of some submerged aquatic plants, the mature pollen grain has a double wall. The vegetative and generative cells are surrounded by a thin delicate wall of unaltered cellulose called the endospore or intine, and a tough resistant outer cuticularized wall composed largely of sporopollenin called the exospore or exine. The exine often bears spines or warts, or is variously sculptured, and the character of the markings is often of value for identifying genus, species, or even cultivar or individual. The spines may be less than a micron in length (spinulus, plural spinuli) referred to as spinulose (scabrate), or longer than a micron (echina, echinae) referred to as echinate. Various terms also describe the sculpturing such as reticulate, a net like appearance consisting of elements (murus, muri) separated from each other by a lumen (plural lumina).

The pollen wall protects the sperm while the pollen grain is moving from the anther to the stigma; it protects the vital genetic material from drying out and solar radiation. The pollen grain surface is covered with waxes and proteins, which are held in place by structures called sculpture elements on the surface of the grain. The outer pollen wall, which prevents the pollen grain from shrinking and crushing the genetic material during desiccation, is composed of two layers. These two layers are the tectum and the foot layer, which is just above the intine. The tectum and foot layer are separated by a region called the columella, which is composed of strengthening rods. The outer wall is constructed with a resistant biopolymer called sporopollenin.

Pollen apertures are regions of the pollen wall that may involve exine thinning or a significant reduction in exine thickness. They allow shrinking and swelling of the grain caused by changes in moisture content. Elongated apertures or furrows in the pollen grain are called colpi (singular: colpus) or sulci (singular: sulcus). Apertures that are more circular are called pores. Colpi, sulci and pores are major features in the identification of classes of pollen. Pollen may be referred to as inaperturate (apertures absent) or aperturate (apertures present). The aperture may have a lid (operculum), hence is described as operculate. However the term inaperturate covers a wide range of morphological types, such as functionally inaperturate (cryptoaperturate) and omniaperturate. Inaperaturate pollen grains often have thin walls, which facilitates pollen tube germination at any position. Terms such as uniaperturate and triaperturate refer to the number of apertures present (one and three respectively).

The orientation of furrows (relative to the original tetrad of microspores) classifies the pollen as sulcate or colpate. Sulcate pollen has a furrow across the middle of what was the outer face when the pollen grain was in its tetrad. If the pollen has only a single sulcus, it is described as monosulcate, has two sulci, as bisulcate, or more, as polysulcate. Colpate pollen has furrows other than across the middle of the outer faces. Eudicots have pollen with three colpi (tricolpate) or with shapes that are evolutionarily derived from tricolpate pollen. The evolutionary trend in plants has been from monosulcate to polycolpate or polyporate pollen.

Pollen Tube

A pollen tube is part of the male gametophyte of seed plants. It acts as a conduit to transport the male gamete cells from the pollen grain, either from the stigma (in flowering plants) to the ovules. In maize, this single cell can grow longer than 12 inches to traverse the length of the pistil.

Pollen tubes were first discovered by Giovanni Battista Amici.

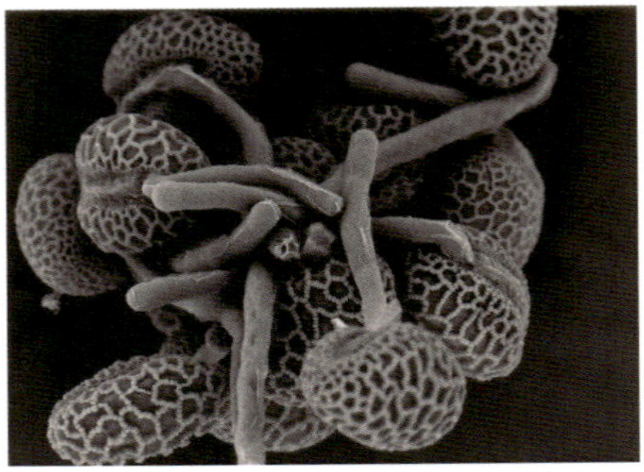

SEM image of pollen tubes growing from *Lily* pollen grains.

The Pollen Tube in Angiosperms

Angiosperm reproduction is a complex process that includes several steps that may vary among species. Each step is a vast procedure in its own right. Pollen is produced by the stamen, the male reproductive organ of the flower. Each pollen grain contains a vegetative cell, and a generative cell that divides to form two sperm cells. The pollen is delivered by the opening of anthers for subsequent pollination, that is, for the transfer of pollen grains to the pistil, the female reproductive organ. Pollination is usually carried out by wind, water or insects. The ovaries hold the ovules that produce the female gamete: the egg cell, which waits in place for fertilization.

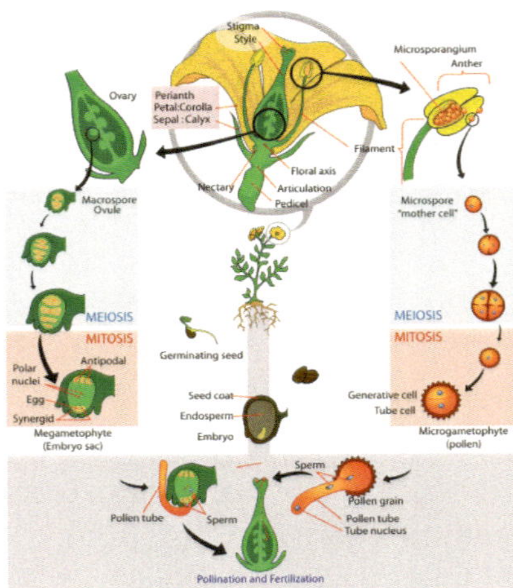

Angiosperm life cycle.

Once a pollen grain settles on a compatible pistil, it may germinate in response to a sugary fluid secreted by the mature stigma of certain plants. Lipids at the surface of the stigma may also stimulate pollen tube growth for compatible pollen. Plants that are self-sterile often inhibit the pollen grains from their own flowers from growing pollen tubes. The presence of multiple grains of pollen has been observed to stimulate quicker pollen tube growth in some plants. The vegetative cell then produces the pollen tube, a tubular protrusion from the pollen grain, which carries the sperm cells within its cytoplasm. The sperm cells are the male gametes that will join with the egg cell and the central cell in double fertilization.

The germinated pollen tube must then drill its way through the nutrient-rich style and curl to the bottom of the ovary to reach the ovule. Once the pollen tube successfully attains an ovule, it delivers the two sperm cells with a burst. One of them fertilizes the egg cell to form an embryo, which will become the future plant. The other one fuses with both polar nuclei of the central cell to form the endosperm, which serves as the embryo's food supply. Finally, the ovary will develop into a fruit and the ovules will develop into seeds.

Pollen Tubes used to Study Cell Physiology

Pollen tubes are an excellent model for the understanding of plant cell behavior. They are easily cultivated in vitro and have a very dynamic cytoskeleton that polymerizes at very high rates, providing the pollen tube with interesting mechanical properties. The pollen tube has an unusual kind of growth; it extends exclusively at its apex. Extending the cell wall only at the tip minimizes friction between the tube and the invaded tissue. This tip growth is performed in a pulsating manner rather than in a steady fashion. Remarkably, the pollen tube's journey through the style often results in depth-to-diameter ratios above 100:1 and up to 1000:1 in certain species. The internal machinery and the external interactions that govern the dynamics of pollen tube growth are far from being fully understood.

Pollen Tube Guidance

Extensive work has been dedicated to comprehend how the pollen tube responds to extracellular guidance signals to achieve fertilization. It is believed that pollen tubes react to a combination of chemical, electrical, and mechanical cues during its journey through the pistil. However, it is not clear how these external cues work or how they are processed internally. Moreover, sensory receptors for any external cue have not been identified yet. Nevertheless, several aspects have already been identified as central in the process of pollen tube growth. The actin filaments in the cytoskeleton, the peculiar cell wall, secretory vesicle dynamics, and the flux of ions, to name a few, are some of the fundamental features readily identified as crucial, but whose role has not yet been completely elucidated.

DNA Repair

During pollen tube growth, DNA damages that arise need to be repaired in order for the male genomic information to be transmitted intact to the next generation. In the plant *Cyrtanthus mackenii*, bicellular mature pollen contains a generative cell and a vegetative cell. Sperm cells are derived by mitosis of the generative cell during pollen tube elongation. The vegetative cell is responsible for pollen tube development. Double-strand breaks in DNA that arise appear to be efficiently repaired in the generative cell, but not in the vegetative cell, during the transport process to the female gametophyte.

References

- "New Genetic Model Predicts Plant Flowering in Different Environments" (Press release). Brown University. January 15, 2009. Retrieved March 20, 2016
- Lloyd, David G.; Barrett, Spencer C. H., eds. (1996). Floral Biology. Studies on Floral Evolution in Animal-Pollinated Plants. ISBN 9781461284949
- Busch, Andrea; Zachgo, Sabine (2009). "Flower symmetry evolution: Towards understanding the abominable mystery of angiosperm radiation". BioEssays. 31 (11): 1181. doi:10.1002/bies.200900081. PMID 19847818
- Physics.org (2012). University of Adelaide. "Flightless parrots, burrowing bats helped parasitic Hades flower". Date Retrieved August 2013
- Taiz, Lincoln; Zeiger, Eduardo; Møller, Ian; Murphy, Angus (2015). Plant Physiology and Development (Sixth ed.). Sunderland, MA: Sinauer Associates, Inc. ISBN 978-1-60535-353-1
- Endress, P. K. (2010). "The evolution of floral biology in basal angiosperms". Philosophical Transactions of the Royal Society B: Biological Sciences. 365 (1539): 411. doi:10.1098/rstb.2009.0228. PMC 2838258
- Purves, William K.; Sadava, David; Orians, Gordon H. (2004). Life: The Science of Biology. Volume III: Plants and Animals. Macmillan. p. 745. Retrieved 2011-01-17
- Schiestl, Florian P. (2010). "The evolution of floral scent and insect chemical communication". Ecology Letters. 13 (5): 643–56. doi:10.1111/j.1461-0248.2010.01451.x. PMID 20337694
- Starr, Cecie; Taggart, Ralph; Evers, Christine; Starr, Lisa (2013). Plant Structure and Function. 4 (13th ed.). Brooks/Cole. p. 517. ISBN 978-1-111-58068-1
- "Capturing Fast Pollen Tube Growth on Camera, Researchers Pin Down Plant Fertilization Process". Retrieved 2018-03-23
- Meneely, Philip (2014). Genetic Analysis: Genes, Genomes, and Networks in Eukaryotes (2 ed.). Oxford University Press. p. 373. ISBN 978-0-19-968126-6
- Yanovsky, M.J.; Kay, S.A. (2002). "Molecular basis of seasonal time measurement in Arabidopsis". Nature. 419 (6904): 308–312. doi:10.1038/nature00996. PMID 12239570.
- Burgess, Jeremy (1985). An Introduction to Plant Cell Development. CUP Archive. p. 55. ISBN 0-521-31611-1. Retrieved 2011-01-17
- Siewertsen, Bjarne. "Hård nyser for allergikere i varm fremtid" (English: Hard sneeze for allergic people in warm future) Danish Meteorological Institute, 18 April 2015. Retrieved: 19 April 2015
- Poltronieri, Palmiro; Hong, Yiguo (2015). Applied Plant Genomics and Biotechnology. Cambridge (UK): Woodhead Publishing. p. 121. ISBN 978-0-08-100068-7

- Chittka, Lars; Walker, Julian (2006). "Do bees like Van Gogh's Sunflowers?" (PDF). Optics & Laser Technology. 38 (4–6): 323. doi:10.1016/j.optlastec.2005.06.020
- Graham, S. W.; Barrett, S. C. H. (1 July 2004). "Phylogenetic reconstruction of the evolution of stylar polymorphisms in Narcissus (Amaryllidaceae)". American Journal of Botany. 91 (7): 1007–1021. doi:10.3732/ajb.91.7.1007. PMID 21653457. Retrieved 25 October 2014
- Parkin, J. (1914). "The Evolution of the Inflorescence". Journal of the Linnean Society of London, Botany. 42 (287): 511. doi:10.1111/j.1095-8339.1914.tb00888.x
- Jones, Hamlyn G. (1992). Plants and Microclimate: A Quantitative Approach to Environmental Plant Physiology. Cambridge University Press. p. 225. ISBN 978-0-521-42524-7
- Jaeger, Katja E.; Wigge, Philip A. "FT Protein Acts as a Long-Range Signal in Arabidopsis". Current Biology. 17 (12): 1050–1054. doi:10.1016/j.cub.2007.05.008
- Barkley, Theodore M.; Brouillet, Luc & Strother, John L. "Asteraceae". Flora of North America. Retrieved 2013-03-04 – via www.eFloras.org. Azcón-Bieto; et al. (2000). Fundamentos de fisiología vegetal. McGraw-Hill/Interamericana de España, SAU. ISBN 84-486-0258-7

3

Flowering Plants: Pollination and Fertilization

Pollination is the transfer of pollen from the male part of a plant to a female part. It enables fertilization and the consequent production of seeds. The aim of this chapter is to explore the different mechanisms of pollination and fertilization in flowers.

Pollination

Pollination is the process by which pollen is transferred to the female reproductive organs of a plant, thereby enabling fertilization to take place. Like all living organisms, seed plants have a single major purpose: to pass their genetic information on to the next generation. The reproductive unit is the seed, and pollination is an essential step in the production of seeds in all spermatophytes (seed plants).

For the process of pollination to be successful, a pollen grain produced by the anther, the male part of a flower, must be transferred to a stigma, the female part of the flower, of a plant of the same species. The process is rather different in angiosperms (flowering plants) from what it is in gymnosperms (other seed plants). In angiosperms, after the pollen grain has landed on the stigma, it creates a pollen tube which grows down the style until it reaches the ovary. Sperm cells from the pollen grain then move along the pollen tube, enter the egg cell through the micropyle and fertilise it, resulting in the production of a seed.

Carpenter bee with pollen collected from Night-blooming cereus

Tip of a tulip stamen covered with pollen grains.

the bee act as pollinator and encourages pollination

Pollination is the process by which pollen is transferred to the female reproductive organs of a plant, thereby enabling fertilization to take place. Like all living organisms, seed plants have a single major purpose: to pass their genetic information on to the next generation. The reproductive unit is the seed, and pollination is an essential step in the production of seeds in all spermatophytes (seed plants).

For the process of pollination to be successful, a pollen grain produced by the anther, the male part of a flower, must be transferred to a stigma, the female part of the flower, of a plant of the same species. The process is rather different in angiosperms (flowering plants) from what it is in gymnosperms (other seed plants). In angiosperms, after the pollen grain has landed on the stigma, it creates a pollen tube which grows down the style until it reaches the ovary. Sperm cells from the pollen grain then move along the pollen tube, enter the egg cell through the micropyle and fertilise it, resulting in the production of a seed.

A successful angiosperm pollen grain (gametophyte) containing the male gametes is transported to the stigma, where it germinates and its pollen tube grows down the style to the ovary. Its two gametes travel down the tube to where the gametophyte(s) containing the female gametes are held within the carpel. One nucleus fuses with the polar bodies to produce the endosperm tissues, and the other with the ovule to produce the embryo Hence the term: "double fertilization".

In gymnosperms, the ovule is not contained in a carpel, but exposed on the surface of a dedicated support organ, such as the scale of a cone, so that the penetration of carpel tissue is unnecessary. Details of the process vary according to the division of gymnosperms in question. Two main modes of fertilization are found in gymnosperms. Cycads and *Ginkgo* have motile sperm that swim directly to the egg inside the ovule, whereas conifers and gnetophytes have sperm that are unable to swim but are conveyed to the egg along a pollen tube.

The study of pollination brings together many disciplines, such as botany, horticulture, entomology, and ecology. The pollination process as an interaction between flower and pollen vector was first addressed by Christian Konrad Sprengel. It is important in horticulture and agriculture, because fruiting is dependent on fertilization: the result of pollination. The study of pollination by insects is known as *anthecology*.

Pollination Process

Pollen germination has three stages; hydration, activation and pollen tube emergence. The pollen grain is severely dehydrated so that its mass is reduced enabling it to be more easily transported from flower to flower. Germination only takes place after rehydration, ensuring that premature germination does not take place in the anther. Hydration allows the plasma membrane of the pollen grain to reform into its normal bilayer organization providing an effective osmotic membrane. Activation involves the development of actin filaments throughout the cytoplasm of the cell, which eventually become concentrated at the point from which the pollen tube will emerge. Hydration and activation continue as the pollen tube begins to grow.

In conifers, the reproductive structures are borne on cones. The cones are either pollen cones (male) or ovulate cones (female), but some species are monoecious and others dioecious. A pollen cone contains hundreds of microsporangia carried on (or borne on) reproductive structures called sporophylls. Spore mother cells in the microsporangia divide by meiosis to form haploid microspores that develop further by two mitotic divisions into immature male gametophytes (pollen grains). The four resulting cells consist of a large tube cell that forms the pollen tube, a generative cell that will produce two sperm by mitosis, and two prothallial cells that degenerate. These cells comprise a very reduced microgametophyte, that is contained within the resistant wall of the pollen grain.

The pollen grains are dispersed by the wind to the female, ovulate cone that is made up of many overlapping scales (sporophylls, and thus megasporophylls), each protecting two ovules, each of which consists of a megasporangium (the nucellus) wrapped in two layers of tissue, the integument and the cupule, that were derived from highly modified branches of ancestral gymnosperms. When a pollen grain lands close enough to the tip of an ovule, it is drawn in through the micropyle (a pore in the integuments covering the tip of the ovule) often by means of a drop of liquid known as a pollination drop. The pollen enters a pollen chamber close to the nucellus, and there it may wait for a year before it germinates and forms a pollen tube that grows through the wall of the megasporangium (=nucellus) where fertilisation takes place. During this time, the megaspore mother cell divides by meiosis to form four haploid cells, three of which degenerate. The surviving one develops as a megaspore and divides repeatedly to form an immature female gametophyte (egg sac). Two or three archegonia containing an egg then develop inside the gametophyte. Meanwhile, in the spring of the second year two sperm cells are produced by mitosis of the body cell of the male gametophyte. The pollen tube elongates and pierces and grows through the megasporangium wall and delivers the sperm cells to the female gametophyte inside. Fertilisation takes place when the nucleus of one of the sperm cells enters the egg cell in the megagametophyte's archegonium.

In flowering plants, the anthers of the flower produce microspores by meiosis. These undergo mitosis to form male gametophytes, each of which contains two haploid cells. Meanwhile, the ovules produce megaspores by meiosis, further division of these form the female gametophytes, which are very strongly reduced, each consisting only of a few cells, one of which is the egg. When a pollen grain adheres to the stigma of a carpel it germinates, developing a pollen tube that grows through the tissues of the style, entering the ovule through the micropyle. When the tube reaches the egg sac, two sperm cells pass through it into the female gametophyte and fertilisation takes place.

Types

Depending on the source of pollen, pollination can be classified into 2 types - Self-pollination and Cross Pollination (Xenogamy). Self Pollination is further divided into Autogamy and Geitonogamy. Depending on agent of Pollination, pollination can be classified into abiotic pollination and biotic pollination.

Self Pollination is the type of Pollination in which pollen grains are transferred from anther to the stigma of the same flower (Autogamy) or pollen grains are transferred from anther to the stigma of different flower of the same plant (Geitonogamy).

Cross Pollination or Xenogamy is the type of pollination in which pollen grains are transferred from anther to the stigma of a different plant.

Abiotic

Abiotic pollination refers to situations where pollination is mediated without the involvement of other organisms. The most common form of abiotic pollination, anemophily, is pollination by wind. Wind pollination is very imprecise, with a minute proportion of pollen grains landing by chance on a suitable receptive stigma, the rest being wasted in the environment. This form of pollination is used by grasses, most conifers, and many deciduous trees. Hydrophily is pollination by water, and occurs in aquatic plants which release their pollen directly into the surrounding water. About 80% of all plant pollination is biotic. In gymnosperms, biotic pollination is generally incidental when it occurs, though some gymnosperms and their pollinators are mutually adapted for pollination. The best-known examples probably are members of the order Cycadales and associated species of beetles. Of the abiotically pollinated species of plant, 98% are anemophilous and 2% hydrophilous, their pollen being transported by water.

It is thought that among angiosperms, entomophily is the primitive state; this is indicated by the vestigial nectaries in the wind-pollinated *Urtica* and other plants, and the presence of fragrances in some of these plants. Of the angiosperms, grasses, sedges, rushes and catkin-bearing plants are in general wind pollinated. Other flowering plants are mostly biotic, the pollen being carried by animal vectors. However a number of plants in multiple families have secondarily adopted wind pollination in contrast to other members of their groups. Some plants are intermediate between the two pollination methods. common heather is regularly pollinated by insects, but produce clouds of pollen and some wind pollination is inevitable, and the hoary plantain is primarily wind pollinated, but is also visited by insects which pollinate it.

Melissodes desponsa covered in pollen

Biotic

Hummingbirds typically feed on red flowers

More commonly, the process of pollination requires pollinators: organisms that carry or move the pollen grains from the anther of one flower to the receptive part of the carpel or pistil (stigma) of another. This is biotic pollination. The various flower traits (and combinations thereof) that differentially attract one type of pollinator or another are known as pollination syndromes. At least 100,000 species of animal, and possibly as many as 200,000, act as pollinators of the estimated 250,000 species of flowering plants in the world. The majority of these pollinators are insects, but about 1,500 species of birds and mammals have been reported to visit flowers and may transfer pollen between them. Besides birds and bats which are the most frequent visitors, these include monkeys, lemurs, squirrels, rodents and possums.

Entomophily, pollination by insects, often occurs on plants that have developed colored petals and a strong scent to attract insects such as, bees, wasps and occasionally ants (Hymenoptera), beetles (Coleoptera), moths and butterflies (Lepidoptera), and flies (Diptera). The existence of insect pollination dates back to the dinosaur era.

In zoophily, pollination is performed by vertebrates such as birds and bats, particularly, hummingbirds, sunbirds, spiderhunters, honeyeaters, and fruit bats. Ornithophily or bird pollination is the pollination of flowering plants by birds. Chiropterophily or bat pollination is the pollination of flowering plants by bats. Plants adapted to use bats or moths as pollinators typically have white petals, strong scent and flower at night, whereas plants that use birds as pollinators tend to produce copious nectar and have red petals.

Insect pollinators such as honey bees (*Apis mellifera*), bumblebees (*Bombus terrestris*), and butterflies (*Thymelicus flavus*) have been observed to engage in flower constancy, which means they are more likely to transfer pollen to other conspecific plants. This can be beneficial for the pollinators, as flower constancy prevents the loss of pollen during interspecific flights and pollinators from clogging stigmas with pollen of other flower species. It also improves the probability that the pollinator will find productive flowers easily accessible and recognisable by familiar clues.

Mechanism

A European honey bee collects nectar, while pollen collects on its body.

Africanized honey bees immersed in Yellow Opuntia engelmannii Cactus Flower Pollen

Diadasia Bee Straddles Cactus Flower Carpels

Pollination can be accomplished by cross-pollination or by self-pollination:

- Cross-pollination, also called *allogamy*, occurs when pollen is delivered from the stamen of one flower to the stigma of a flower on another plant of the same species. Plants adapted for cross-pollination have several mechanisms to prevent self-pollination; the reproductive organs may be arranged in such a way that self-fertilisation is unlikely, or the stamens and carpels may mature at different times.

- Self-pollination occurs when pollen from one flower pollinates the same flower or other flowers of the same individual. It is thought to have evolved under conditions when pollinators were not reliable vectors for pollen transport, and is most often seen in short-lived annual species and plants that colonize new locations. Self-pollination may include *autogamy*, where pollen is transferred to the female part of the same flower; or *geitonogamy*, when pollen is transferred to another flower on the same plant. Plants adapted to self-fertilize often have similar stamen and carpel lengths. Plants that can pollinate themselves and produce viable offspring are called self-fertile. Plants that cannot fertilize themselves are called self-sterile, a condition which mandates cross-pollination for the production of offspring.

- *Cleistogamy*: is self-pollination that occurs before the flower opens. The pollen is released from the anther within the flower or the pollen on the anther grows a tube down the style to the ovules. It is a type of sexual breeding, in contrast to asexual systems such as apomixis. Some *cleistogamous* flowers never open, in contrast to *chasmogamous* flowers that open and are then pollinated. Cleistogamous flowers are by necessity found on self-compatible or self-fertile plants. Although certain orchids and grasses are entirely cleistogamous, other plants resort to this strategy under adverse conditions. Often there may be a mixture of both cleistogamous and chasmogamous flowers, sometimes on different parts of the plant and sometimes in mixed inflorescences. The ground bean produces cleistogamous flowers below ground, and mixed cleistogamous and chasmogamous flowers above.

Geranium incanum, like most geraniums and pelargoniums, sheds its anthers, sometimes its stamens as well, as a barrier to self-pollination. This young flower is about to open its anthers, but has not yet fully developed its pistil.

These *Geranium incanum* flowers have opened their anthers, but not yet their stigmas. Note the change of colour that signals to pollinators that it is ready for visits.

This *Geranium incanum* flower has shed its stamens, and deployed the tips of its pistil without accepting pollen from its own anthers. (It might of course still receive pollen from younger flowers on the same plant.)

An estimated 48.7% of plant species are either dioecious or self-incompatible obligate out-crossers. It is also estimated that about 42% of flowering plants exhibit a mixed mating system in nature. In the most common kind of mixed mating system, individual plants produce a single type of flower and fruits may contain self-pollinated, out-crossed or a mixture of progeny types.

Pollination also requires consideration of pollenizers. The terms "pollinator" and "pollenizer" are often confused: a *pollinator* is the agent that moves the pollen, whether it be bees, flies, bats, moths, or birds; a *pollenizer* is the plant that serves as the pollen source for other plants. Some plants are *self-compatible* (*self-fertile*) and can pollinate and fertilize themselves. Other plants have chemical or physical barriers to self-pollination.

In agriculture and horticulture pollination management, a good pollenizer is a plant that provides compatible, viable and plentiful pollen and blooms at the same time as the plant that is to be pollinated or has pollen that can be stored and used when needed to pollinate the desired flowers. Hybridization is effective pollination between flowers of different species, or between different breeding lines or populations.

Peaches are considered self-fertile because a commercial crop can be produced without cross-pollination, though cross-pollination usually gives a better crop. Apples are considered self-incompatible, because a commercial crop must be cross-pollinated. Many commercial fruit tree varieties are grafted clones, genetically identical. An orchard block of apples of one variety is genetically a single plant. Many growers now consider this a mistake. One means of correcting this mistake is to graft a limb of an appropriate pollenizer (generally a variety of crabapple) every six trees or so.

The wasp *Mischocyttarus* rotundicollis transporting pollen grains of *Schinus terebinthifolius*

Pollen Vectors

Biotic pollen vectors are animals, usually insects, but also reptiles, birds, mammals, and sundry others, that routinely transport pollen and play a role in pollination. This is usually as a result of their activities when visiting plants for feeding, breeding or shelter. The pollen adheres to the vector's body parts such as face, legs, mouthparts, hair, feathers, and moist spots; depending on the particular vector. Such transport is vital to the pollination of many plant species.

Any kind of animal that often visits or encounters flowers is likely to be a pollen vector to some extent. For example, a crab spider that stops at one flower for a time and then moves on, might carry pollen incidentally, but most pollen vectors of significant interest are those that routinely visit the flowers for some functional activity. They might feed on pollen, or plant organs, or on plant secretions such as nectar, and carry out acts of pollination on the way. Many plants bear flowers that favour certain types of pollinator over all others. This need not always be an effective strategy, because some flowers that are of such a shape that they favor pollinators that pass by their anthers and stigmata on the way to the nectar, may get robbed by ants that are small enough to bypass the normal channels, or by short-tongued bees that bite through the bases of deep corolla tubes to extract nectar at the end opposite to the anthers and stigma. Some pollinator species can show huge variation in pollination effectiveness because their ability to carry pollen is impacted by some morphological trait. This is the case in the white-lined sphinx moth, in which short-tongued morphs collect pollen on their heads but long-tongued morphs do not carry any pollen. Some flowers have specialized mechanisms to trap pollinators to increase effectiveness. Other flowers will attract pollinators by odor. For example, bee species such as *Euglossa cordata* are attracted to orchids this way, and it has been suggested that the bees will become intoxicated during these visits to the orchid flowers, which last up to 90 minutes. However, in general, plants that rely on pollen vectors tend to be adapted to their particular type of vector, for example day-pollinated species tend to be brightly coloured, but if they are pollinated largely by birds or specialist mammals, they tend to be larger and have larger nectar rewards than species that are strictly insect-pollinated. They also tend to spread their rewards over longer periods, having long flowering seasons; their specialist pollinators would be likely to starve if the pollination season were too short.

As for the types of pollinators, reptile pollinators are known, but they form a minority in most ecological situations. They are most frequent and most ecologically significant in island systems, where insect and sometimes also bird populations may be unstable and less species-rich. Adaptation to a lack of animal food and of predation pressure, might therefore favour reptiles becoming more herbivorous and more inclined to feed on pollen and nectar. Most species of lizards in the families that seem to be significant in pollination seem to carry pollen only incidentally, especially the larger species such as Varanidae and Iguanidae, but especially several species of the Gekkonidae are active pollinators, and so is at least one species of the Lacertidae, *Podarcis lilfordi*, which pollinates various species, but in particular is the major pollinator of *Euphorbia dendroides* on various Mediterranean islands.

Mammals are not generally thought of as pollinators, but some rodents, bats and marsupials are significant pollinators and some even specialise in such activities. In South Africa certain species of *Protea* (in particular *Protea humiflora*, *P. amplexicaulis*, *P. subulifolia*, *P. decurrens* and *P. cordata*) are adapted to pollination by rodents (particularly Cape Spiny Mouse, *Acomys subspinosus*) and elephant shrews (*Elephantulus* species). The flowers are borne near the ground, are yeasty smelling, not colourful, and sunbirds reject the nectar with its high xylose content. The mice apparently can digest the xylose and they eat large quantities of the pollen. In Australia pollination by flying, gliding and earthbound mammals has been demonstrated. Examples of pollen vectors include many species of wasps, that transport pollen of many plant species, being potential or even efficient pollinators.

Evolution of Plant/pollinator Interactions

The first fossil record for abiotic pollination is from fern-like plants in the late Carboniferous period. Gymnosperms show evidence for biotic pollination as early as the Triassic period. Many fossilized pollen grains show characteristics similar to the biotically dispersed pollen today. Furthermore, the gut contents, wing structures, and mouthpart morphologies of fossilized beetles and flies suggest that they acted as early pollinators. The association between beetles and angiosperms during the early Cretaceous period led to parallel radiations of angiosperms and insects into the late Cretaceous. The evolution of nectaries in late Cretaceous flowers signals the beginning of the mutualism between hymenopterans and angiosperms.

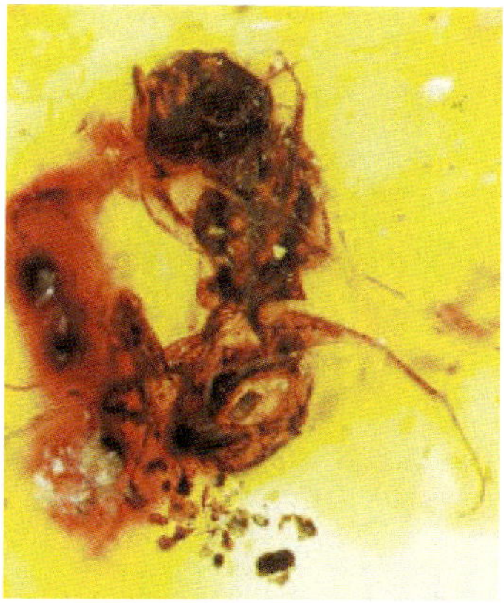

Melittosphex burmensis, the oldest bee fossil, from the Cretaceous

Bees provide a good example of the mutualism that exists between hymenopterans and angiosperms. Flowers provide bees with nectar (an energy source) and pollen (a source of protein). When bees go from flower to flower collecting pollen they are also depositing pollen grains onto the flowers, thus pollinating them. While pollen and nectar, in most cases, are the most notable reward attained from flowers, bees also visit flowers for other resources such as oil, fragrance, resin and even waxes. It has been estimated that bees originated with the origin or diversification of angiosperms. In addition, cases of coevolution between bee species and flowering plants have been illustrated by specialized adaptations. For example, long legs are selected for in *Rediviva neliana*, a bee that collects oil from *Diascia capsularis*, which have long spur lengths that are selected for in order to deposit pollen on the oil-collecting bee, which in turn selects for even longer legs in *R. neliana* and again longer spur length in *D. capsularis* is selected for, thus, continually driving each other's evolution.

Environmental Impacts

Loss of pollinators, also known as Pollinator decline (of which colony collapse disorder is perhaps the most well known) has been noticed in recent years. Observed losses would have significant economic impacts. Possible explanations for pollinator decline include habitat destruction, pesticide, parasitism/diseases, climate change and others, and many researchers believe it is the synergistic effects of these factors which are ultimately detrimental to pollinator populations.

The Structure of Plant–pollinator Networks

Wild pollinators often visit a large number of plant species and plants are visited by a large number of pollinator species. All these relations together form a network of interactions between plants and pollinators. Surprising similarities were found in the structure of networks consisting out of the interactions between plants and pollinators. This structure was found to be similar in very different ecosystems on different continents, consisting of entirely different species.

The structure of plant-pollinator networks may have large consequences for the way in which pollinator communities respond to increasingly harsh conditions. Mathematical models, examining the consequences of this network structure for the stability of pollinator communities suggest that the specific way in which plant-pollinator networks are organized minimizes competition between pollinators and may even lead to strong indirect facilitation between pollinators when conditions are harsh. This means that pollinator species together can survive under harsh conditions. But it also means that pollinator species collapse simultaneously when conditions pass a critical point. This simultaneous collapse occurs, because pollinator species depend on each other when surviving under difficult conditions.

Such a community-wide collapse, involving many pollinator species, can occur suddenly when increasingly harsh conditions pass a critical point and recovery from such a collapse might not be easy. The improvement in conditions needed for pollinators to recover, could be substantially larger than the improvement needed to return to conditions at which the pollinator community collapsed.

Cross-pollination

Xenogamy is the transfer of pollen grains from the anther to the stigma of a different plant. This is the only type of pollination which during pollination brings genetically different types of pollen grains to the stigma.

The term xenogamy (along with geitonogamy and autogamy) was first suggested by Kerner in 1876. Cross-pollination involves the transfer of pollen grains from the flower of one plant to the stigma of the flower of another plant.

The main characteristics which facilitate cross-pollination are:

- Herkogamy: Flowers possess some mechanical barrier on their stigmatic surface to avoid self-pollination, e.g. presence of gynostegium and pollinia in *Calotropis*.
- Dichogamy: Pollen and stigma of the flower mature at different times to avoid self-pollination.
- Self-incompatibility: In same plants, the mature pollen fall on the receptive stigma of the same flower but fail to bring about self-pollination.

- Male sterility: The pollen grains of some plants are not functional. Such plants set seeds only after cross-pollination.
- Dioecism: Cross-pollination always occurs when the plants are unisexual and dioecious, i.e., male and female flowers occur on separate plants, e.g., papaya, some cucurbits, etc.
- Heterostyly: The flowers of some plants have different lengths of stamens and styles so that self-pollination is not possible, e.g., *Primula*, *Linum*, etc.

Self-pollination

One type of automatic self-pollination occurs in the orchid *Ophrys apifera*. One of the two pollinia bends itself towards the stigma.

Self-pollination is when pollen from the same plant arrives at the stigma of a flower (in flowering plants) or at the ovule (in gymnosperms). There are two types of self-pollination: in autogamy, pollen is transferred to the stigma of the same flower; in geitonogamy, pollen is transferred from the anther of one flower to the stigma of another flower on the same flowering plant, or from microsporangium to ovule within a single (monoecious) gymnosperm. Some plants have mechanisms that ensure autogamy, such as flowers that do not open (cleistogamy), or stamens that move to come into contact with the stigma. The term selfing that is often used as a synonym, is not limited to self-pollination, but also applies to other types of self-fertilization.

Occurrence

Few plants self-pollinate without the aid of pollen vectors (such as wind or insects). The mechanism is seen most often in some legumes such as peanuts. In another legume, soybeans, the flowers open and remain receptive to insect cross pollination during the day. If this is not accomplished, the flowers self-pollinate as they are closing. Among other plants that can self-pollinate are many kinds of orchids, peas, sunflowers and tridax. Most of the self-pollinating plants have small, relatively inconspicuous flowers that shed pollen directly onto the stigma, sometimes even before the bud opens.

Self-pollinated plants expend less energy in the production of pollinator attractants and can grow in areas where the kinds of insects or other animals that might visit them are absent or very scarce—as in the Arctic or at high elevations.

Self-pollination limits the variety of progeny and may depress plant vigor. However, self-pollination can be advantageous, allowing plants to spread beyond the range of suitable pollinators or produce offspring in areas where pollinator populations have been greatly reduced or are naturally variable.

Pollination can also be accomplished by cross-pollination. Cross-pollination is the transfer of pollen, by wind or animals such as insects and birds, from the anther to the stigma of flowers on separate plants.

Types of Flowers that Self Pollinate

Both hermaphrodite and monoecious species have the potential for self-pollination leading to self-fertilization unless there is a mechanism to avoid it. Eighty percent of all flowering plants are hermaphroditic, meaning they contain both sexes in the same flower, while 5 percent of plant species are monoecious. The remaining 15% would therefore be dioecious (each plant unisexual). Plants that self-pollinate include several types of orchids, and sunflowers. Dandelions are also capable of self-pollination as well as cross-pollination.

Advantages of Self-pollination

There are several advantages for self-pollinating flowers. Firstly, if a given genotype is well-suited for an environment, self-pollination helps to keep this trait stable in the species. Not being dependent on pollinating agents allows self-pollination to occur when bees and wind are nowhere to be found. Self-pollination can be an advantage when the number of flowers are small or widely spaced. During self-pollination, the pollen grains are not transmitted from one flower to another. As a result, the wastage of pollen is less. Also, it does not depend on any external carrier. The plant that develops in that way of pollination cannot make changes in their characters and so the features of a species can be maintained. This way, a species can maintain purity in it.

Disadvantages of Self-pollination

The disadvantages of self-pollination come from a lack of variation that allows no adaptation to the changing environment or potential pathogen attack. Self-pollination can lead to inbreeding depression caused by expression of deleterious recessive mutations, or to the reduced health of the species, due to the breeding of related specimens. This is why many flowers that could potentially self-pollinate have a built-in mechanism to avoid it, or make it second choice at best. Genetic defects in self-pollinating plants cannot be eliminated by genetic recombination and offspring can only avoid inheriting the deleterious attributes through a chance mutation arising in a gamete.

Mixed Mating

About 42% of flowering plants exhibit a mixed mating system in nature. In the most common kind of system, individual plants produce a single flower type and fruits may contain self-pollinated, out-crossed or a mixture of progeny types. Another mixed mating system is referred to as dimorphic cleistogamy. In this system a single plant produces both open, potentially out-crossed and closed, obligately self-pollinated cleistogamous flowers.

Self-pollinating Species

The evolutionary shift from outcrossing to self-fertilization is one of the most common evolutionary transitions in plants. About 10-15% of flowering plants are predominantly self-fertilizing. A few well-studied examples of self-pollinating species are described below.

Paphiopedilum Parishii

Self-pollination in the slipper orchid *Paphiopedilum parishii* occurs when the anther changes from a solid to a liquid state and directly contacts the stigma surface without the aid of any pollinating agent.

Holcoglossum Amesianum

The tree-living orchid *Holcoglossum amesianum* has a type of self-pollination mechanism in which the bisexual flower turns its anther against gravity through 360° in order to insert pollen into its own stigma cavity---without the aid of any pollinating agent or medium. This type of self-pollination appears to be an adaptation to the windless, drought conditions that are present when flowering occurs, at a time when insects are scarce. Without pollinators for outcrossing, the necessity of ensuring reproductive success appears to outweigh potential adverse effects of inbreeding. Such an adaptation may be widespread among species in similar environments.

Caulokaempferia Coenobialis

In the Chinese herb *Caulokaempferia coenobialis* a film of pollen is transported from the anther (pollen sacs) by an oily emulsion that slides sideways along the flower's style and into the individual's own stigma. The lateral flow of the film of pollen along the style appears to be due solely to the spreading properties of the oily emulsion and not to gravity. This strategy may have evolved to cope with a scarcity of pollinators in the extremely shady and humid habitats of *C. coenobialis*.

Capsella Rubella

Capsella rubella (Red Shepard's purse) is a self-pollinating species that became self-compatible 50,000 to 100,000 years ago, indicating that self-pollination is an evolutionary adaptation that can persist over many generations. Its out-crossing progenitor was identified as *Capsella grandiflora*.

Arabidopsis Thaliana

Arabidopsis thaliana is a predominantly self-pollinating plant with an out-crossing rate in the wild estimated at less than 0.3%. A study suggested that self-pollination evolved roughly a million years ago or more.

Bulbophyllum Bicoloratum

Self-pollination in the Madagascan orchid Bulbophyllum bicoloratum occurs by virtue of a rostellum that may have regained its stigmatic function as part of the distal median stigmatic lobe.

Possible Long-term Benefit of Meiosis

Meiosis followed by self-pollination produces little overall genetic variation. This raises the question of how meiosis in self-pollinating plants is adaptively maintained over extended periods [i.e. for roughly a million years or more, as in the case of *A.thaliana*] in preference to a less complicated and less costly asexual ameiotic process for producing progeny. An adaptive benefit of meiosis that may explain its long-term maintenance in self-pollinating plants is efficient recombinational repair of DNA damage. This benefit can be realized at each generation (even when genetic variation is not produced).

Geitonogamy

Geitonogamy is a type of self-pollination. Geitonogamous pollination is sometimes distinguished from the fertilizations that can result from it, geitonogamy. If a plant is self-incompatible, geitonogamy can reduce seed production.

Geitonogamy is when pollen is exported using a vector (pollinator or wind) out of one flower but only to another flower on the same plant. It is a form of self-fertilization.

In flowering plants, pollen is transferred from a flower to another flower on the same plant, and in animal pollinated systems this is accomplished by a pollinator visiting multiple flowers on the same plant. Geitonogamy is also possible within species that are wind-pollinated, and may actually be a quite common source of self-fertilized seeds in self-compatible species. It also occurs in monoecious gymnosperms. Although geitonogamy is functionally cross-pollination involving a pollinating agent, genetically it is similar to autogamy since the pollen grains come from the same plant.

Monoecious plants like maize show geitonogamy. Geitonogamy is not possible for strictly dioecious plants.

Cleistogamy

Cleistogamy is a type of automatic self-pollination of certain plants that can propagate by using non-opening, self-pollinating flowers. Especially well known in peanuts, peas, and beans, this behavior is most widespread in the grass family. However, the largest genus of cleistogamous plants is *Viola*.

The more common opposite of cleistogamy, or "closed marriage," is called chasmogamy, or "open marriage." Virtually all plants that produce cleistogamous flowers also produce chasmogamous ones. The principal advantage of cleistogamy is that it requires fewer plant resources to produce seeds than does chasmogamy, because development of petals, nectar and large amounts of pollen is not required. This efficiency makes cleistogamy particularly useful for seed production on unfavorable sites or adverse conditions. *Impatiens capensis,* for example, has been observed to produce only cleistogamous flowers after being severely damaged by grazing and to maintain populations on unfavorable sites with only cleistogamous flowers. The obvious disadvantage of cleistogamy is that self-fertilization occurs, which may suppress the creation of genetically superior plants.

For genetically modified (GM) rapeseed, researchers hoping to minimise the admixture of GM and non-GM crops are attempting to use cleistogamy to prevent gene flow. However, preliminary results from Co-Extra, a current project within the EU research program, show that although cleistogamy reduces gene flow, it is not at the moment a consistently reliable tool for biocontainment; due to a certain instability of the cleistogamous trait, some flowers may open and release genetically modified pollen.

Chasmogamy

Chasmogamous (a) and cleistogamous (b) flowers of *Viola pubescens*. Arrows point to structure.

Chasmogamy, is a plant reproductive mechanism in which pollination occurs in chasmogamous flowers. Chasmogamous flowers are commonly showy with open petals encircling exposed reproductive parts. Chasmogamous stems from Greek for "open marriage", named after the open arrangement of floral structures. Once chasmogamous flowers have reached maturity, they unfurl and their stamens and/or style are made available for pollination. Although some plant species possess self-fertilizing chasmogamous flowers, most chasmogamous flowers are cross-pollinated by biotic (e.g. insects) or abiotic (e.g. wind) agents.

Chasmogamous flowers that are cross-pollinated have the advantage of sexual reproduction between two different parents, resulting in sexual recombination and genetically distinct seeds. These distinct seeds increase genetic diversity of the plant population and reduce inbreeding depression and deleterious alleles. Seeds produced by chasmogamous flowers that are cross-pollinated, may also exhibit hybrid vigor (heterosis) and increased fitness.

To promote cross-pollination, chasmogamous flowers often have strikingly colored petals and nectar guides/nectaries to attract and reward pollinators. However, these attractive floral organs can be disadvantageous as they are energetically costly to produce and require a surplus of resources. Pollination of chasmogamous flowers is also dependent upon the availability of pollinating agents. Thus, chasmogamous flowers generally develop when resources like light, water, nutrients, and pollinators are high.

In contrast to chasmogamous flowers, are minute, bud-like cleistogamous ("closed marriage") flowers, and pollination of cleistogamous flowers is cleistogamy. Unlike chasmogamous flowers, cleistogamous flowers remain mechanically sealed throughout the entirety of their development and reproduction. The closed morphology of cleistogamous flowers hinders them from exposing their reproductive organs and forces self-pollination. Without the need for pollinating agents, cleistogamous flowers lack nectar and elaborate petals, making them much less costly to produce than chasmogamous flowers and developmentally favored in suboptimal conditions.

In addition to chasmogamy and cleistogamy, many plant species possess a mixed breeding system that utilizes both chasmogamous and cleistogamous flowers. This breeding system has been referred to under multiple names including true cleistogamy, dimorphic cleistogamy, and the chasmogamous/cleistogamous mixed breeding system. In this mixed breeding system, species produce both chasmogamous and cleistogamous flowers on the same plant. Chasmogamous and cleistogamous flowers may exhibit spatial separation and develop simultaneously at different morphological locations, or the two flowers may be temporally separated and develop at different times during the flowering season. Rarely, the two flowers have also been documented as being both spatially and temporally separated. The presence of both flower types allows mixed breeding system species to cross-pollinate and self-fertilize. This leads to increases in genetic diversity and assures reproductive success in variable environmental and pollinator conditions.

Hand-pollination

Hand pollination, also known as mechanical pollination is a technique that can be used to pollinate plants when natural or open pollination is either undesirable or insufficient. This method of pollination is done by manually transferring pollen from the stamen of one plant to the pistil of another. This is often done with a cotton swab or small brush, but can also be done by removing the petals from a male flower and brushing it against the stigmas of female flowers, or by simply shaking flowers in the case of bisexual flowers, such as tomatoes. Common reasons for choosing this method include the lack of pollinators, keeping control of cross-pollination between varieties grown together, and creating specific hybrids. Examples of this are vanilla plants, which are transported to areas where its natural pollinator doesn't exist, or plants grown in greenhouses, urban areas, or with a cover to control pests, where natural pollinators cannot reach them. Pollinator decline and the concentrated pollination needs of monoculture can also be a factor. However, these are not the only reasons, and variable techniques for hand-pollination have arisen for many specialty crops. For instance, hand-pollination is used with date palms to avoid wasting space and energy growing sufficient male plants for adequate natural pollination. Because of the level of labor involved, hand-pollination is only an option on a small scale, used chiefly by small market gardeners and owners of individual plants. On large-scale operations, such as field crops, orchards, or commercial seed production, honeybees or other pollinators are a more efficient approach to pollination management. Despite this, hand-pollination is a fairly widespread practice. Pears grown in Hanyuan County, China have been hand-pollinated since the 1980s, because they can't be pollinated with other varieties that have different flowering times; also, lice infestation requires the use of many insecticide sprays, which causes local beekeepers to refuse to lend beehives.

Hand-pollination of two gourd blossoms

Self Fertilization

Autogamy, or self-fertilization, refers to the fusion of two gametes that come from one individual. Autogamy is predominantly observed in the form of self-pollination, a reproductive mechanism employed by many flowering plants. However, species of protists have also been observed using autogamy as a means of reproduction. Flowering plants engage in autogamy regularly, while the protists that engage in autogamy only do so in stressful environments.

Self-pollination in Flowering Plants

Self-pollination is an example of autogamy that occurs in flowering plants. Self-pollination occurs when the sperm in the pollen from the stamen of a plant goes to the carpels of that same plant and fertilizes the egg cell present. Self-pollination can either be done completely autogamously or geitonogamously. In the former, the egg and sperm cells that united came from the same flower. In the latter, the sperm and egg cells can come from a different flower on the same plant. While the latter method does blur the lines between autogamous self-fertilization and normal sexual reproduction, it is still considered autogamous self-fertilization.

Self-pollination can lead to inbreeding depression due to expression of deleterious recessive mutations. Meiosis followed by self-pollination results in little genetic variation, raising the question of how meiosis in self-pollinating plants is adaptively maintained over an extended period in preference to a less complicated and less costly asexual ameiotic process for producing progeny. For instance, *Arabidopsis thaliana* is a predominantly self-pollinating plant that has an outcrossing rate in the wild estimated at less than 0.3%, and self-pollination appears to have evolved roughly a million years ago or more. An adaptive benefit of meiosis that may explain its long-term maintenance in self-pollinating plants is efficient recombinational repair of DNA damage.

Advantages of Autogamy

There are several advantages for the self-fertilization observed in flowering plants and protists. In flowering plants, it is important for some plants not to be dependent on pollinating agents that other plants rely for fertilization. This is unusual, however, considering that many plant species have evolved to become incompatible with their own gametes. While these species would not be well served by having autogamous self-fertilization as a reproductive mechanism, other species, which do not have self-incompatibility, would benefit from autogamy. Protists have the advantage of diversifying their modes of reproduction. This is useful for a multitude of reasons. First, if there is an unfavorable change in the environment that puts the ability to deliver offspring at risk, then it is advantageous for an organism to have autogamy at its disposal. In other organisms, it is seen that genetic diversity arising from sexual reproduction is maintained by changes in the environment that favor certain genotypes over others. Aside from extreme circumstances, it is possible that this form of reproduction gives rise to a genotype in the offspring that will increase fitness in the environment. This is due to the nature of the genetic degeneration and remodeling intrinsic to autogamy in unicellular organisms. Thus, autogamous behavior may become advantageous to have if an individual wanted to ensure offspring viability and survival. This advantage also applies to flowering plants. However, it is important to note that this change has not shown to produce a progeny with more fitness in unicellular organisms. It is possible that the nutrition deprived state of the parent cells before autogamy created a barrier for producing offspring that could thrive in those same stressful environments.

Disadvantages of Autogamy

In flowering plants, autogamy has the disadvantage of producing low genetic diversity in the species that use it as the predominant mode of reproduction. This leaves those species particularly susceptible to pathogens and viruses that can harm it. In addition, the foraminiferans that use autogamy have shown to produce substantially smaller progeny as a result. This indicates that since it is generally an emergency survival mechanism for unicellular species, the mechanism does not have the nutritional resources that would be provided by the organism if it were undergoing binary fission.

Double Fertilization

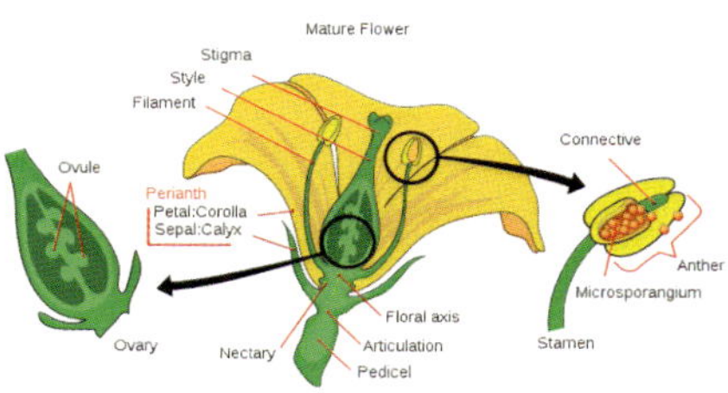

The parts of a flower

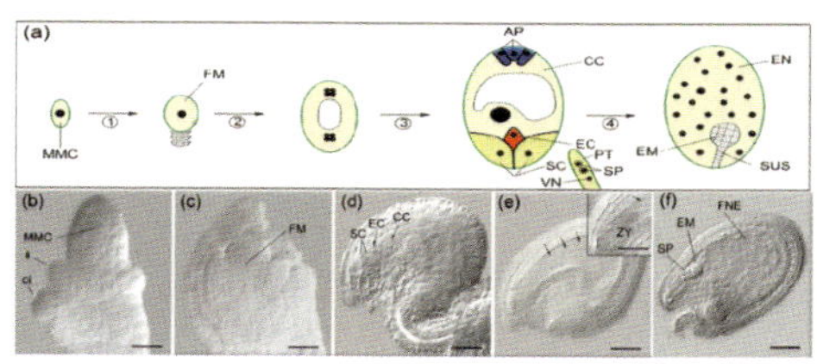

Double fertilization in *Arabidopsis*

Double fertilization is a complex fertilization mechanism of flowering plants (angiosperms). This process involves the joining of a female gametophyte (megagametophyte, also called the embryo sac) with two male gametes (sperm). It begins when a pollen grain adheres to the stigma of the carpel, the female reproductive structure of a flower. The pollen grain then takes in moisture and begins to germinate, forming a pollen tube that extends down toward the ovary through the style. The tip of the pollen tube then enters the ovary and penetrates through the micropyle opening in the ovule. The pollen tube proceeds to release the two sperm in the megagametophyte.

The ovules are 8 in number and arranged in the form of 3+2+3(from top to bottom) i.e 2 synergids & 1 egg cell, 2 polar central cells, 3 antipodal cells. One sperm fertilizes the egg cell and the other sperm combines with the two polar nuclei of the large central cell of the megagametophyte. The haploid sperm and haploid egg combine to form a diploid zygote,the process being called syngamy , while the other sperm and the two haploid polar nuclei of the large central cell of the megagametophyte form a triploid nucleus (triple fusion). Some plants may form polyploid nuclei. The large cell of the gametophyte will then develop into the endosperm, a nutrient-rich tissue which provides nourishment to the developing embryo. The ovary, surrounding the ovules, develops into the fruit, which protects the seeds and may function to disperse them.

The two central cell maternal nuclei (polar nuclei) that contribute to the endosperm, arise by mitosis from the same single meiotic product that gave rise to the egg. The maternal contribution to the genetic constitution of the triploid endosperm is double that of the embryo.

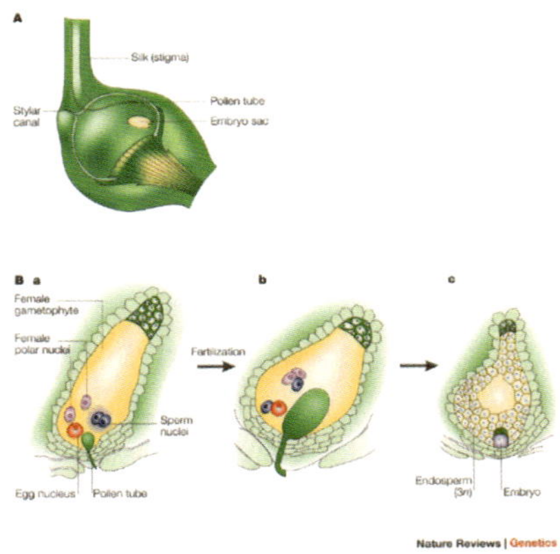

Double fertilization

In a study conducted in 2008 of the plant *Arabidopsis thaliana*, the migration of male nuclei inside the female gamete, in fusion with the female nuclei, has been documented for the first time using in vivo imaging. Some of the genes involved in the migration and fusion process have also been determined.

Evidence of double fertilization in Gnetales, which are non-flowering seed plants, has been reported.

In Vitro Double Fertilization

In vitro double fertilization is often used to study the molecular interactions as well as other aspects of gamete fusion in flowering plants. One of the major obstacles in developing an in vitro double fertilization between male and female gametes is the confinement of the sperm in the pollen tube and the egg in the embryo sac. A controlled fusion of the egg and sperm has already been achieved with poppy plants. Pollen germination, pollen tube entry, and double fertilization processes have all been observed to proceed normally. In fact, this technique has already been used to obtain seeds in various flowering plants and was named "test-tube fertilization".

Related Structures and Functions

Megagametophyte

The female gametophyte, the megagametophyte, that participates in double fertilization in angiosperms which is haploid is called the embryo sac. This develops within an ovule, enclosed by the ovary at the base of a carpel. Surrounding the megagametophyte are (one or) two integuments, which form an opening called the micropyle. The megagametophyte, which is usually haploid, originates from the (usually diploid) megaspore mother cell, also called the megasporocyte. The next sequence of events varies, depending on the particular species, but in most species, the following events occur. The megasporocyte undergoes a meiotic cell division, producing four haploid megaspores. Only one of the four resulting megaspores survives. This megaspore undergoes three rounds of mitotic division, resulting in seven cells with eight haploid nuclei (the central cell has two nuclei, called the polar nuclei). The lower end of the embryo sac consists of the haploid egg cell positioned in the middle of two other haploid cells, called synergids. The synergids function in the attraction and guidance of the pollen tube to the megagametophyte through the micropyle. At the upper end of the megagametophyte are three antipodal cells.

Microgametophyte

The male gametophytes, or microgametophytes, that participate in double fertilization are contained within pollen grains. They develop within the microsporangia, or pollen sacs, of the anthers on the stamens. Each microsporangium contains diploid microspore mother cells, or microsporocytes. Each microsporocyte undergoes meiosis, forming four haploid microspores, each of which can eventually develop into a pollen grain. A microspore undergoes mitosis and cytokinesis in order to produce two separate cells, the generative cell and the tube cell. These two cells in addition to the spore wall make up an immature pollen grain. As the male gametophyte matures, the generative cell passes into the tube cell, and the generative cell undergoes mitosis, producing two sperm cells. Once the pollen grain has matured, the anthers break open, releasing the pollen. The pollen is carried to the pistil of another flower, by wind or animal pollinators, and deposited on the stigma. As the pollen grain germinates, the tube cell produces the pollen tube, which elongates and extends down the long style of the carpel and into the ovary, where its sperm cells are released in the megagametophyte. Double fertilization proceeds from here.

References

- Mauseth, James D. Botany: An Introduction to Plant Biology. Publisher: Jones & Bartlett, 2008 ISBN 978-0-7637-5345-0

- Ackerman, J. D. (2000-03-01). "Abiotic pollen and pollination: Ecological, functional, and evolutionary perspectives". Plant Systematics and Evolution. 222 (1-4): 167–185. doi:10.1007/BF00984101
- Sühs, R.B.; Somavilla, A.; Putzke, J.; Köhler, A. (2009). "Pollen vector wasps (Hymenoptera, Vespidae) of Schinus terebinthifolius Raddi (Anacardiaceae)". Brazilian Journal of Biosciences. 7 (2): 138–143
- "First ever record of insect pollination from 100 million years ago". ScienceDaily. Retrieved 2015-10-20
- Darwin, Charles. More Letters of Charles Darwin, Volume 2. Echo Library. p. 668. ISBN 978-1-4068-0482-9. Retrieved 25 February 2012
- Friedman, Jannice; Barrett, Spencer C. H. (2009-06-01). "Wind of change: new insights on the ecology and evolution of pollination and mating in wind-pollinated plants". Annals of Botany. 103 (9): 1515–1527. doi:10.1093/aob/mcp035. PMC 2701749
- Campbell, Neil A.; Reece, Jane B. (2002). Biology (6th edition). Pearson Education. pp. 600–612. ISBN 978-0-201-75054-6
- Runions, C. John; Owens, John N. (1999). "Sexual reproduction of interior spruce (Pinaceae). I. Pollen germination to archegonial maturation". International Journal of Plant Sciences. 160 (4): 631–640. doi:10.1086/314170
- Rai, Nagendra; Rai, Mathura (2006). Heterosis breeding in vegetable crops. New India Publishing. ISBN 978-81-89422-03-5. Retrieved July 5, 2011
- Lee, J. J. and McEnery, M. E. (1970), Autogamy in Allogromia laticollaris (Foraminifera). The Journal of Protozoology, 17: 184–195
- Abrol, Dharam P. (2012). "Non Bee Pollinators-Plant Interaction". Pollination Biology. Chapter 9: 265–310. doi:10.1007/978-94-007-1942-2_9
- Rodríguez-Gironés, Miguel A.; Santamaría, Luis (2004). "Why are so many bird flowers red?". PLoS Biology. 2 (10): e306. doi:10.1371/journal.pbio.0020350. PMC 521733. PMID 15486585
- Cronk, J. K.; Fennessy, M. Siobhan (2001). Wetland plants: biology and ecology. Boca Raton, Fla.: Lewis Publishers. p. 166. ISBN 1-56670-372-7
- Fan, Xu-Li; Barrett, Spencer C. H.; Lin, Hua; Chen, Ling-Ling; Zhou, Xiang; Gao, Jiang-Yun (2012-10-01). "Rain pollination provides reproductive assurance in a deceptive orchid". Annals of Botany. 110 (5): 953–958. doi:10.1093/aob/mcs1654
- Baskin, Carol C.; Baskin, Jerry M. (2001). Seeds: Ecology, Biogeography, and Evolution of Dormancy and Germination. Elsevier. p. 215. ISBN 978-0-12-080263-0
- McLaughlin, Chris (2010). "5. Pure Pollination". The Complete Idiot's Guide to Heirloom Vegetables. Penguin. p. 56. ISBN 978-1-61564-052-2. Retrieved July 5, 2011

4

Flower Species

Flowering plants form the most diverse group among land plants with over two hundred thousand known species. Such plants are widely grown for aesthetic appeal. Poinsettias, chrysanthemums, orchids and azaleas are commonly cultivated flowering plants. The topics elaborated in this chapter will help in gaining a better perspective about the plants that are of significance in floriculture.

Poinsettia

The poinsettia (*Euphorbia pulcherrima*) is a commercially important plant species of the diverse spurge family (Euphorbiaceae). The species is indigenous to Mexico. It is particularly well known for its red and green foliage and is widely used in Christmas floral displays. It derives its common English name from Joel Roberts Poinsett, the first United States Minister to Mexico, who introduced the plant to the US in 1825.

Description

Euphorbia pulcherrima is a shrub or small tree, typically reaching a height of 0.6–4 metres (2–13 ft). The plant bears dark green dentate leaves that measure 7–16 centimetres (2.8–6.3 in) in length. The colored bracts—which are most often flaming red but can be orange, pale green, cream, pink, white, or marbled—are often mistaken for flower petals because of their groupings and colors, but are actually leaves. The colors of the bracts are created through photoperiodism, meaning that they require darkness (12 hours at a time for at least five days in a row) to change color. At the same time, the plants require abundant light during the day for the brightest color.

The flowers of the poinsettia are unassuming and do not attract pollinators. They are grouped within small yellow structures found in the center of each leaf bunch, and are called cyathia.

The poinsettia is native to Mexico. It is found in the wild in deciduous tropical forests at moderate elevations from southern Sinaloa down the entire Pacific coast of Mexico to Chiapas and Guatemala. It is also found in the interior in the hot, seasonally dry forests of Guerrero, Oaxaca, and Chiapas. Reports of *E. pulcherrima* growing in the wild in Nicaragua and Costa Rica have yet to be confirmed by botanists.

Cultivation

A poinsettia that was stressed and reflowered after growing in bonsai form

Poinsettia varieties on sale in England

A close-up photo of a poinsettia (*Euphorbia pulcherrima*)

The poinsettia has been cultivated in Egypt since the 1860s, when it was brought from Mexico during the Egyptian campaign. It is called *bent el consul*, "the consul's daughter", referring to the U.S. ambassador Joel Poinsett.

There are over 100 cultivated varieties of poinsettia.

In areas outside its natural environment, it is commonly grown as an indoor plant where it prefers good morning sun, then shade in the hotter part of the day. Contrary to popular belief, flowering poinsettias can be kept outside, even during winter, as long as they are kept frost-free. It is widely grown and very popular in subtropical climates such as Australia, Rwanda and Malta.

The plant requires a daily period of uninterrupted long, dark nights followed by bright sunny days for around two months in autumn in order to encourage it to develop colored bracts. Any incidental light during these nights (from a nearby television set, from under a door frame, even from passing cars or street lights) hampers bract production. Commercial production of poinsettia has been done by placing them inside a greenhouse and covering the latter completely to imitate the natural biological situation.

To produce extra axillary buds that are necessary for plants containing multiple flowers, a phytoplasma infection—whose symptoms include the proliferation of axillary buds—is used. The discovery of the role phytoplasmas play in the growth of axillary buds is credited to Ing-Ming Lee of the USDA Agricultural Research Service.

Diseases

Poinsettias are susceptible to several diseases, mostly fungal, but also bacterial and parasitic.

Toxicity Claims

In the United States and perhaps elsewhere, there is a common misconception that the poinsettia is highly toxic. This misconception was spread by a 1919 urban legend of a two-year-old child dying after consuming a poinsettia leaf.

While the sap and latex of many plants of the spurge genus are indeed toxic, the poinsettia's toxicity is relatively mild. Its latex can cause an allergic reaction in sensitive individuals. It is also mildly irritating to the skin or stomach and may sometimes cause diarrhea and vomiting if eaten. Sap introduced into the human eye may cause temporary blindness.

An *American Journal of Emergency Medicine* study of 22,793 cases reported to the American Association of Poison Control Centers showed no fatalities, and furthermore that a strong majority of poinsettia exposures are accidental, involve children, and usually do not result in any type of medical treatment. POISINDEX, a major source for poison control centers, says a 50 lb (23 kg) child would have to eat 500 bracts (poinsettia leaves) to accumulate levels of toxins found to be harmful in experiments. An Ohio State University study showed no problems even with extremely large doses.

Pansy

The garden pansy is a type of large-flowered hybrid plant cultivated as a garden flower. It is derived by hybridization from several species in the section *Melanium* ("the pansies") of the genus *Viola*, particularly *Viola tricolor*, a wildflower of Europe and western Asia known as heartsease. Some of these hybrids are referred to as *Viola* × *wittrockiana* Gams ex Nauenb. & Buttler. For simplicity, the older name *Viola tricolor* var. *hortensis* is often used.

The garden pansy flower is 5 to 8 centimetres (2 to 3 in) in diameter and has two slightly overlapping upper petals, two side petals, and a single bottom petal with a slight beard emanating from the flower's center. These petals are usually white or yellow, purplish, or blue. The plant may grow to 23 cm (9 in) in height, and prefers sun to varying degrees and well-draining soils.

Names and Terminology

Pansy displaying the two upper overlapping petals, the two side petals, and the single bottom petal

Pansies growing at the edge of the pavement

English common names, such as "pansy", "viola" and "violet" may be used interchangeably. One possible distinction is that plants considered to be "pansies" are classified in *Viola* sect. *Melanium*, and have four petals pointing upwards (the two side petals point upwards), and only one pointing down, whereas those considered to be "violets" are classified in *Viola* sect. *Viola*, and have two petals pointing up and three pointing down. Another possible distinction is made by the American Violet Society – the International Cultivar Registration Authority for the genus *Viola*. It divides cultivated varieties (cultivars) in *Viola* sect. *Melanium* into four subgroups: B1 – pansies, B2 – violas, B3 – violettas and B4 – cornuta hybrids. On this classification, modern "pansies" differ from the other three subgroups by possessing a well-defined "blotch" or "eye" in the middle of the flower.

Modern horticulturalists tend to use the term "pansy" for those multi-coloured large-flowered hybrids that are grown for bedding purposes every year, while "viola" is usually reserved for smaller, more delicate annuals and perennials.

Name in Different Languages, Countries

In the United States, pansies have been colloquially referred to as "football flowers" because of the Milwaukee "Football" or soccer decorations that use white chrysanthemums and black pansies to create a soccer ball (no other flower gets quite as black as a pansy).

In Scandinavia, Scotland, and German-speaking countries, the pansy (or its wild parent *Viola tricolor*) is or was known as the "stepmother"; the name was accompanied by an aitiological tale about a selfish stepmother, told to children while the teller plucked off corresponding parts of the blossom to fit the plot.

In Italy the pansy is known as *flammola* (little flame).

In Hungary it is known as *árvácska* (small orphan).

In Israel, the pansy is known as "Amnon and Tamar", after the tragic story involving three of King David's children.

Cultivation

Modern horticulturists have developed a wide range of pansy flower colors and bicolors including yellow, gold, orange, purple, violet, red, white, and even near-black (very dark purple). Pansies typically display large showy face markings.

Pansies produced for the bedding market

Plants grow well in sunny or partially sunny positions in well-draining soils. Pansies are perennial, but normally grown as biennials or annuals because of their leggy growth. The first year plant produces greenery, and bears flowers and seeds in its second year of growth. Afterwards, the plant dies like an annual. Because of selective human breeding, most garden pansies bloom the first year, some in as little as nine weeks after sowing.

After flowering, a seed capsule matures, eventually opening as seen here.

Pansies are purchased as six-packs or "flats" (USA) of young plants from garden centers and planted directly into the garden soil. Plants will grow up to 23 cm (9 in) in height with flowers measuring 5 to 8 cm (2 to 3 in) in diameter, though smaller and larger flowering cultivars are available.

Pansies in a British Garden Centre

Pansies are winter hardy in zones 4–8. They can survive light freezes and short periods of snow cover, but, in areas with prolonged snow cover, a covering of a dry winter mulch is recommended. In warmer climates, zones 9-11, pansies can bloom over the winter, and are often planted in the fall. In warmer zones, pansies may re-seed themselves and return the next year. They are not very heat-tolerant; warm temperatures inhibit blooming and hot muggy air causes rot and death. In colder zones, pansies may not survive without snow cover or protection (mulch) from extreme cold or periods of freezing and thawing. They perform best in zones with moderate temperatures, and equal amounts of mild rainfall and sunshine.

Pansies, for best growth, are watered thoroughly about once a week, depending on climate and rainfall. The plant should never be over-watered. To maximize blooming, plant foods are used about every other week, depending on the type of food used. Regular deadheading can extend the blooming period.

Pests and Diseases

Aphid and her young

Aphids

Aphids, which can spread the cucumber mosaic virus, sometimes feed on pansies. Infestations are treated with a spray of diluted soft soap (2 ounces per gallon) or insecticides.

Leaf Spot

Leaf spot (*Ramularia deflectens*) is a fungal infection. Symptoms include dark spots on leaf margins followed by a white web covering the leaves. It is associated with cool damp springs.

Downy Mildew

Pansy downy mildew is caused by the fungus-like organism *Peronospora violae*, which produces purple-brown leaf spots, often with encircling yellowing, that have an accompanying grey mold on the leaf underside. It can severely weaken or kill affected plants.

Powdery Mildew

A disease caused by one or more species of fungus in the Erysiphaceae family. Symptoms include violet-gray powder on fringes and underside of leaves. It is caused by stagnant air and can be limited but not necessarily eliminated by spraying (especially leaf undersides).

A pansy exhibiting the flower's morphology: two large petals overlapping at the top, two side petals, a lower petal with slight indentation, and beards at the center

Slugs and Snails

Slugs and snails feed on the foliage. Sometimes sharp, gritty sand or a top-dressing of chipped bark is used by gardeners to limit damage.

Stem Rot

Stem rot, also known as pansy sickness, is a soil-borne fungus and a possible hazard with unsterilized animal manure. The plant may collapse without warning in the middle of the season. The foliage will flag and lose color. Flowers will fade and shrivel prematurely. Stem will snap at the soil line if tugged slightly. The plant is probably a total loss unless tufted. The treatment of stem rot includes the use of fungicides such as Cheshunt or Benomyl, which are used prior to planting. Infected plants are destroyed (burned) to prevent the spread of the pathogen to other plants.

Cucumber Mosaic Virus

The cucumber mosaic virus is transmitted by aphids. Pansies with the virus have fine yellow veining on young leaves, stunted growth and anomalous flowers. The virus can lay dormant, affect the entire plant and be passed to next generations and to other species. Prevention is key: purchases should consist entirely of healthy plants, and pH-balanced soil should be used which is neither too damp nor too dry. The soil should have balanced amounts of nitrogen, phosphate and potash. Other diseases which may weaken the plant should be eliminated.

Tulip

Tulips (*Tulipa*) form a genus of spring-blooming perennial herbaceous bulbiferous geophytes (having bulbs as storage organs). The flowers are usually large, showy and brightly coloured, generally red, yellow, or white. They often have a different coloured blotch at the base of the tepals (petals and sepals, collectively), internally. Because of a degree of variability within the populations, and a long history of cultivation, classification has been complex and controversial. The tulip is a member of the Liliaceae (lily) family, along with 14 other genera, where it is most closely related to *Amana*, *Erythronium* and *Gagea* in the tribe Lilieae. There are about 75 species, and these are divided between four subgenera. The name "tulip" is thought to be derived from a Persian word for turban, which it may have been thought to resemble. Tulips originally were found in a band stretching from Southern Europe to Central Asia, but since the seventeenth century have become widely naturalised and cultivated. In their natural state they are adapted to steppes and mountainous areas with temperate climates. Flowering in the spring, they become dormant in the summer once the flowers and leaves die back, emerging above ground as a shoot from the underground bulb in early spring.

Netherlands, during the time of the Dutch Golden Age and Tulip mania, an infection of tulip bulbs by the tulip breaking virus created variegated patterns in the tulip flowers that were much admired and valued. This phenomenon was referred to as "broken". While tulips had probably been cultivated in Asia, they did not come to the attention of the west, when western diplomats to the Ottoman court observed and reported on them. They were rapidly introduced into Europe and cultivated and became a frenzied commodity during Tulip mania. Tulips were frequently depicted in paintings of the Dutch Golden Age, and have become associated with the Netherlands, the major producer for world markets, ever since.

Breeding programs have produced thousands of hybrid and cultivars in addition to the original species (known in horticulture as botanical tulips). They are popular throughout the world, both as ornamental garden plants and as cut flowers.

Description

Tulip Morphology

Tulipa (tulips) is a genus of spring-blooming perennial herbaceous bulbiferous geophytes, dying back after flowering to an underground storage bulb. Depending on the species, tulip plants can be between 4 inches (10 cm) and 28 inches (71 cm) high.

Flowers: The tulip's flowers are usually large and are actinomorphic (radially symmetric) and hermaphrodite (contain both male (androecium) and female (gynoecium) characteristics), generally erect, or more rarely pendulous, and are arranged more usually as a single terminal flower, or when pluriflor as two to three (e.g. *Tulipa turkestanica*), but up to four, flowers on the end of a floriferous stem (scape), which is single arising from amongst the basal leaf rosette. In structure, the flower is generally cup or star shaped. As with other members of Liliaceae the perianth is undifferentiated (perigonium) and biseriate (two whorled), formed from six free (i.e. apotepalous) caducous tepals arranged into two separate whorls of three parts (trimerous) each. The two whorls represent three petals and three sepals, but are termed tepals because they are nearly identical. The tepals are usually petaloid (petal like), being brightly coloured, but each whorl may be different, or have different coloured blotches at their bases, forming darker colouration on the interior surface. Tulip flowers come in a wide variety of colours, except pure blue (several tulips with "blue" in the name have a faint violet hue), and have absent nectaries.

Bulbs, showing tunic and scales

Star shaped flower of *Tulipa clusiana* with three sepals and three petals, forming six identical tepals

Cup shaped flower of *Tulipa orphanidea*

Androecium: The flowers have six distinct, basifixed introrse stamens arranged in two whorls of three, which vary in length and may be glabrous or hairy. The filaments are shorter than the tepals, and dilated towards their base.

Gynoecium: The style is short or absent and each stigma has three distinct lobes, and the ovaries are superior, with three chambers.

Fruit: The tulip's fruit is a globose or ellipsoid capsule with a leathery covering and an ellipsoid to globe shape. Each capsule contains numerous flat, disc-shaped seeds in two rows per chamber. These light to dark brown seeds have very thin seed coats and endosperm that does not normally fill the entire seed.

Leaves: Tulip stems have few leaves. Larger species tend to have multiple leaves. Plants typically have two to six leaves, some species up to 12. The tulip's leaf is cauline (born on a stem), strap-shaped, with a waxy coating, and the leaves are alternate (alternately arranged on the stem), diminishing in size as the further up the stem. These fleshy blades are often bluish-green in colour.The bulbs are truncated basally and elongated towards the apex, on the outside they are covered with a thin tunic, that is lined with hairs on the inside, but occasionally glabrous.

Phytochemistry

Tulipanin is an anthocyanin found in tulips. It is the 3-rutinoside of delphinidin. The chemical compounds named tuliposides and tulipalins can also be found in tulips and are responsible for allergies. Tulipalin A, or α-methylene-γ-butyrolactone, is a common allergen, generated by hydrolysis of the glucoside tuliposide A. It induces a dermatitis that is mostly occupational and affects tulip bulb sorters and florists who cut the stems and leaves. Tulipanin A and B are toxic to horses, cats and dogs.

Distribution and Habitat

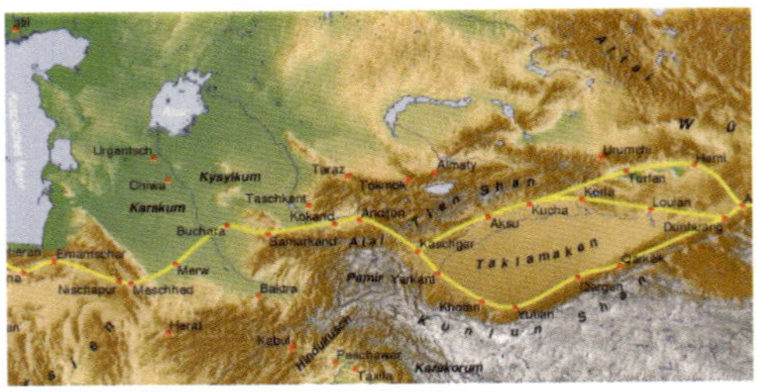

Eastern end of the tulip range from Turkmenistan on the eastern shore of the Caspian Sea to the Pamir-Alai and Tien-Shan mountains

Tulips are mainly distributed along a band corresponding to Latitude 40° north, from south-east of Europe (Ukraine, Russia) and Turkey in the west, through the Levant (Syria, Israel, Palestinian Territories, Lebanon and Jordan). From there it extends eastwards through Jerevan, (Armenia) and Baku (Azerbaijan) and on the eastern shore of the Caspian Sea through Turkmenistan, Bukhara, Samarkand and Tashkent (Uzbekistan), to the eastern end of the range in the Pamir-Alai and Tien-Shan mountains in Central Asia, which form the centre of diversity. Further to the east, *Tulipa* is found in the western Himalayas, southern Siberia, inner Mongolia, and as far as the northwest of China. While authorities have stated that no tulips west of the Balkans are native, subsequent identification of *Tulipa sylvestris* subsp. *australis* as a native of the Iberian peninsula and adjacent North Africa shows that this may be a simplification. In addition to these regions in the west tulips have been identified in Greece, Cyprus and the Balkans. In the south, Iran marks its furthest extent, while the northern limit is the Ukraine. Although tulips are also throughout most of the Mediterranean and Europe, these regions do not form part of the natural distribution. Tulips were brought to Europe by travelers and merchants from Anatolia and Central Asia for cultivation, from where they escaped and naturalised. For instance, less than a half of those species found in Turkey are actually native. These have been referred to as neo-tulipae.

Tulips are indigenous to mountainous areas with temperate climates, where they are a common element of steppe and winter-rain Mediterranean vegetation. They thrive in climates with long, cool springs and dry summers. Tulips are most commonly found in meadows, steppes and chaparral, but also introduced in fields, orchards, roadsides and abandoned gardens.

Ecology

Variegation produced by the tulip breaking virus

Botrytis tulipae is a major fungal disease affecting tulips, causing cell death and eventually the rotting of the plant. Other pathogens include anthracnose, bacterial soft rot, blight caused by *Sclerotium rolfsii*, bulb nematodes, other rots including blue molds, black molds and mushy rot.

The fungus *Trichoderma viride* can infect tulips, producing dried leaf tips and reduced growth, although symptoms are usually mild and only present on bulbs growing in glasshouses.

Variegated tulips admired during the Dutch tulipomania gained their delicately feathered patterns from an infection with the tulip breaking virus, a mosaic virus that was carried by the green peach aphid, *Myzus persicae*. While the virus produces fantastically streaked flowers, it also weakens plants and reduces the number of offsets produced. Tulips affected by mosaic virus are called "broken"; while such plants can occasionally revert to a plain or solid colouring, they will remain infected and have to be destroyed. Today the virus is almost eradicated from tulip growers' fields. The multicoloured patterns or modern varieties result from breeding; they normally have solid, not feathered borders between the colours. Tulips need a period of cool dormancy, known as vernalisation. Tulip bulbs imported to warm-winter areas are often planted in autumn to be treated as annuals.

Cultivation

Propagation

The Netherlands are the world's main producer of commercial tulip plants, producing as many as 3 billion bulbs annually, the majority for export.

Tulips can be propagated through bulb offsets, seeds or micropropagation. Offsets and tissue culture methods are means of asexual propagation for producing genetic clones of the parent plant, which maintains cultivar genetic integrity. Seeds are most often used to propagate species and subspecies or to create new hybrids. Many tulip species can cross-pollinate with each other, and when wild tulip populations overlap geographically with other tulip species or subspecies, they often hybridise and create mixed populations. Most commercial tulip cultivars are complex hybrids, and often sterile.

Tulip pistil surrounded by stamens.

Tulip stamen: note the pollen grains.

The reproductive organs of a tulip

Offsets require a year or more of growth before plants are large enough to flower. Tulips grown from seeds often need five to eight years before plants are of flowering size. Commercial growers usually harvest the tulip bulbs in late summer and grade them into sizes; bulbs large enough to flower are sorted and sold, while smaller bulbs are sorted into sizes and replanted for sale in the future.

Horticultural Classification

'Yonina', a division 6 cultivar

'Texas Flame', a division 10 cultivar

In horticulture, tulips are divided up into fifteen groups (Divisions) mostly based on flower morphology and plant size.

- Div. 1: Single early – with cup-shaped single flowers, no larger than 8 cm across (3 inches). They bloom early to mid season. Growing 15 to 45 cm tall.
- Div. 2: Double early – with fully double flowers, bowl shaped to 8 cm across. Plants typically grow from 30–40 cm tall.

- Div. 3: Triumph – single, cup shaped flowers up to 6 cm wide. Plants grow 35–60 cm tall and bloom mid to late season.
- Div. 4: Darwin hybrid – single flowers are ovoid in shape and up to 8 cm wide. Plants grow 50–70 cm tall and bloom mid to late season. This group should not be confused with older Darwin tulips, which belong in the Single Late Group below.
- Div. 5: Single late – cup or goblet-shaded flowers up to 8 cm wide, some plants produce multi-flowering stems. Plants grow 45–75 cm tall and bloom late season.
- Div. 6: Lily-flowered – the flowers possess a distinct narrow 'waist' with pointed and reflexed petals. Previously included with the old Darwins, only becoming a group in their own right in 1958.
- Div. 7: Fringed (Crispa).
- Div. 8: Viridiflora.
- Div. 9: Rembrandt.
- Div. 10: Parrot.
- Div. 11: Double late – Large, heavy blooms. They range from 18 to 22 inches tall.
- Div. 12: Kaufmanniana – Waterlily tulip. Medium-large creamy yellow flowers marked red on the outside and yellow at the center. Stems 6 in. tall.
- Div. 13: Fosteriana (Emperor).
- Div. 14: Greigii – Scarlet flowers 6 in. across, on 10 in. stems. Foliage mottled with brown.
- Div. 15: Species (Botanical).
- Div. 16: Multiflowering – not an official division, these tulips belong in the first 15 divisions but are often listed separately because they have multiple blooms per bulb.

They may also be classified by their flowering season:

- Early flowering: Single Early Tulips, Double Early Tulips, Greigii Tulips, Kaufmanniana Tulips, Fosteriana Tulips, Species tulips.
- Mid-season flowering: Darwin Hybrid Tulips, Triumph Tulips, Parrot Tulips.
- Late season flowering: Single Late Tulips, Double Late Tulips, Viridiflora Tulips, Lily-flowering Tulips, Fringed Tulips, Rembrandt Tulips.

Species Tulips

The terms "species tulips" and "botanical tulips" refer to wild species in contrast to hybridised varieties. As a group they have been described as being less ostentatious but more reliably vigorous as they age.

Neo-tulipae

A number of names are based on naturalised garden tulips, and are usually referred to as neo-tulipae. These are often difficult to trace back to their original cultivar, and in some cases have been occurring in the wild for many centuries. The history of naturalisation is unknown, but populations are usually associated with agricultural practices and are possibly linked to saffron cultivation . Some neo-tulipae have been brought into cultivation, and are often offered as botanical tulips. These cultivated plants can be classified into two Cultivar Groups: 'Grengiolensis Group', with picotee tepals, and the 'Didieri Group' with unicolourous tepals.

Horticulture

Tulip bulbs are typically planted around late summer and fall, in well-drained soils. Tulips should be planted 4 to 6 inches (10 to 15 cm) apart from each other. The recommended hole depth is 4 to 8 inches (10 to 20 cm) deep, and is measured from the top of the bulb to the surface. Therefore, larger tulip bulbs would require deeper holes. Species tulips are normally planted deeper.

Rose

A rose is a woody perennial flowering plant of the genus *Rosa,* in the family Rosaceae, or the flower it bears. There are over a hundred species and thousands of cultivars. They form a group of plants that can be erect shrubs, climbing or trailing with stems that are often armed with sharp prickles. Flowers vary in size and shape and are usually large and showy, in colours ranging from white through yellows and reds. Most species are native to Asia, with smaller numbers native to Europe, North America, and north-western Africa. Species, cultivars and hybrids are all widely grown for their beauty and often are fragrant. Roses have acquired cultural significance in many societies. Rose plants range in size from compact, miniature roses, to climbers that can reach seven meters in height. Different species hybridize easily, and this has been used in the development of the wide range of garden roses.

The name *rose* comes from French, itself from Latin *rosa*, which was perhaps borrowed from Oscan, from Greek, itself borrowed from Old Persian, related to Avestan, Sogdian, Parthian.

Botany

Longitudinal section through a developing rose hip

Exterior view of rose buds

Rose leaflets

Size can be as small as a thumbnail

The leaves are borne alternately on the stem. In most species they are 5 to 15 centimetres (2.0 to 5.9 in) long, pinnate, with (3–) 5–9 (–13) leaflets and basal stipules; the leaflets usually have a serrated margin, and often a few small prickles on the underside of the stem. Most roses are deciduous but a few (particularly from South east Asia) are evergreen or nearly so.

The flowers of most species have five petals, with the exception of *Rosa sericea*, which usually has only four. Each petal is divided into two distinct lobes and is usually white or pink, though in a few species yellow or red. Beneath the petals are five sepals (or in the case of some *Rosa sericea*, four). These may be long enough to be visible when viewed from above and appear as green points alternating with the rounded petals.

There are multiple superior ovaries that develop into achenes. Roses are insect-pollinated in nature.

The hybrid garden rose "Amber Flush"

The aggregate fruit of the rose is a berry-like structure called a rose hip. Many of the domestic cultivars do not produce hips, as the flowers are so tightly petalled that they do not provide access for pollination. The hips of most species are red, but a few (e.g. *Rosa pimpinellifolia*) have dark purple to black hips. Each hip comprises an outer fleshy layer, the hypanthium, which contains 5–160 "seeds" (technically dry single-seeded fruits called achenes) embedded in a matrix of fine, but stiff, hairs. Rose hips of some species, especially the dog rose (*Rosa canina*) and rugosa rose (*Rosa rugosa*), are very rich in vitamin C, among the richest sources of any plant. The hips are eaten by fruit-eating birds such as thrushes and waxwings, which then disperse the seeds in their droppings. Some birds, particularly finches, also eat the seeds.

Rose thorns are actually prickles – outgrowths of the epidermis.

While the sharp objects along a rose stem are commonly called "thorns", they are technically prickles—outgrowths of the epidermis (the outer layer of tissue of the stem). (True thorns, as produced by e.g. *Citrus* or *Pyracantha*, are modified stems, which always originate at a node and which have nodes and internodes along the length of the thorn itself.) Rose prickles are typically sickle-shaped hooks, which aid the rose in hanging onto other vegetation when growing over it. Some species such as *Rosa rugosa* and *Rosa pimpinellifolia* have densely packed straight prickles, probably an adaptation to reduce browsing by animals, but also possibly an adaptation to trap wind-blown sand and so reduce erosion and protect their roots (both of these species grow naturally on coastal sand dunes). Despite the presence of prickles, roses are frequently browsed by deer. A few species of roses have only vestigial prickles that have no points.

Species

The genus *Rosa* is subdivided into four subgenera:

- *Hulthemia* (formerly *Simplicifoliae*, meaning "with single leaves") containing one or two species from southwest Asia, *R. persica* and *Rosa berberifolia* which are the only roses without compound leaves or stipules.
- *Hesperrhodos* contains *Rosa minutifolia* and *Rosa stellata*, from North America.
- *Platyrhodon* with one species from east Asia, *Rosa roxburghii* (also known as the chestnut rose).
- *Rosa* (the type subgenus, sometimes incorrectly called *Eurosa*) containing all the other roses. This subgenus is subdivided into 11 sections.
    - *Banksianae* – white and yellow flowered roses from China.
    - *Bracteatae* – three species, two from China and one from India.
    - *Caninae* – pink and white flowered species from Asia, Europe and North Africa.
    - *Carolinae* – white, pink, and bright pink flowered species all from North America.
    - *Chinensis* – white, pink, yellow, red and mixed-color roses from China and Burma.
    - *Gallicanae* – pink to crimson and striped flowered roses from western Asia and Europe.
    - *Gymnocarpae* – one species in western North America (*Rosa gymnocarpa*), others in east Asia.
    - *Laevigatae* – a single white flowered species from China
    - *Pimpinellifoliae* – white, pink, bright yellow, mauve and striped roses from Asia and Europe.
    - *Rosa* (syn. sect. *Cinnamomeae*) – white, pink, lilac, mulberry and red roses from everywhere but North Africa.
    - *Synstylae* – white, pink, and crimson flowered roses from all areas.

Rosa gallica *Evêque*, painted by Redouté

Uses

Roses are best known as ornamental plants grown for their flowers in the garden and sometimes indoors. They have been also used for commercial perfumery and commercial cut flower crops. Some are used as landscape plants, for hedging and for other utilitarian purposes such as game cover and slope stabilization. They also have minor medicinal uses.

Ornamental Plants

The majority of ornamental roses are hybrids that were bred for their flowers. A few, mostly species roses are grown for attractive or scented foliage (such as *Rosa glauca* and *Rosa rubiginosa*), ornamental thorns (such as *Rosa sericea*) or for their showy fruit (such as *Rosa moyesii*).

Hybrid tea rose cultivar 'Mrs. Herbert Stevens'

Ornamental roses have been cultivated for millennia, with the earliest known cultivation known to date from at least 500 BC in Mediterranean countries, Persia, and China. Many thousands of rose hybrids and cultivars have been bred and selected for garden use as flowering plants. Most are double-flowered with many or all of the stamens having mutated into additional petals.

The Empress Josephine of France patronized the development of rose breeding at her gardens at Malmaison. As long ago as 1840 a collection numbering over one thousand different cultivars, varieties and species was possible when a rosarium was planted by Loddiges nursery for Abney Park Cemetery, an early Victorian garden cemetery and arboretum in England.

Cut Flowers

Bouquet of pink roses

Roses are a popular crop for both domestic and commercial cut flowers. Generally they are harvested and cut when in bud, and held in refrigerated conditions until ready for display at their point of sale.

In temperate climates, cut roses are often grown in glasshouses, and in warmer countries they may also be grown under cover in order to ensure that the flowers are not damaged by weather and that pest and disease control can be carried out effectively. Significant quantities are grown in some tropical countries, and these are shipped by air to markets across the world.

Some kind of roses are artificially coloured using dyed water, like rainbow roses.

Perfume

Rose perfumes are made from rose oil (also called attar of roses), which is a mixture of volatile essential oils obtained by steam distilling the crushed petals of roses. An associated product is rose water which is used for cooking, cosmetics, medicine and religious practices. The production technique originated in Persia and then spread through Arabia and India, and more recently into eastern Europe. In Bulgaria, Iran and Germany, damask roses (*Rosa* × *damascena* 'Trigintipetala') are used. In other parts of the world *Rosa* × *centifolia* is commonly used. The oil is transparent pale yellow or yellow-grey in colour. 'Rose Absolute' is solvent-extracted with hexane and produces a darker oil, dark yellow to orange in colour. The weight of oil extracted is about one three-thousandth to one six-thousandth of the weight of the flowers; for example, about two thousand flowers are required to produce one gram of oil.

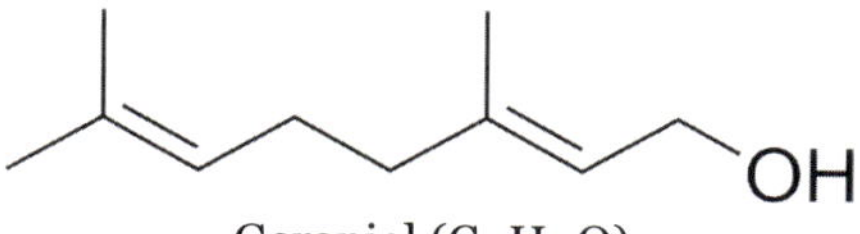

Geraniol ($C_{10}H_{18}O$)

The main constituents of attar of roses are the fragrant alcohols geraniol and L-citronellol and rose camphor, an odorless solid composed of alkanes, which separates from rose oil. β-Damascenone is also a significant contributor to the scent.

Food and Drink

Rose hips are occasionally made into jam, jelly, marmalade, and soup or are brewed for tea, primarily for their high vitamin C content. They are also pressed and filtered to make rose hip syrup. Rose hips are also used to produce rose hip seed oil, which is used in skin products and some makeup products.

Rosa canina hips

Rose water has a very distinctive flavour and is used heavily in Middle Eastern, Persian, and South Asian cuisine—especially in sweets such as barfi, baklava, halva, gulab jamun, gumdrops, kanafeh, nougat, and Turkish delight.

Gulab jamun made with rose water

Rose jams are partially common in Iran

Rose petals or flower buds are sometimes used to flavour ordinary tea, or combined with other herbs to make herbal teas.

In France, there is much use of rose syrup, most commonly made from an extract of rose petals. In the Indian subcontinent, Rooh Afza, a concentrated squash made with roses, is popular, as are rose-flavoured frozen desserts such as ice cream and kulfi.

Rose flowers are used as food, also usually as flavouring or to add their scent to food. Other minor uses include candied rose petals.

Rose creams (rose-flavoured fondant covered in chocolate, often topped with a crystallised rose petal) are a traditional English confectionery widely available from numerous producers in the UK.

Under the American Federal Food, Drug, and Cosmetic Act, there are only certain *Rosa* species, varieties, and parts are on the *Generally Recognized as Safe* lists.

- Rose absolute: *Rosa alba* L., *Rosa centifolia* L., *Rosa damascena* Mill., *Rosa gallica* L., and vars. of these spp
- Rose (otto of roses, attar of roses): Ditto
- Rose buds: Ditto
- Rose flowers: Ditto
- Rose fruit (hips):Ditto
- Rose leaves: *Rosa spp*

Medicine

The rose hip, usually from *R. canina*, is used as a minor source of vitamin C. The fruits of many species have significant levels of vitamins and have been used as a food supplement. Many roses have been used in herbal and folk medicines. *Rosa chinensis* has long been used in Chinese traditional medicine. This and other species have been used for stomach problems, and are being investigated for controlling cancer growth. In pre-modern medicine, diarrhodon is a name given to various compounds in which red roses are an ingredient.

Culture

Art

Roses are a favored subject in art and appear in portraits, illustrations, on stamps, as ornaments or as architectural elements. The Luxembourg-born Belgian artist and botanist Pierre-Joseph Redouté is known for his detailed watercolours of flowers, particularly roses.

Codex Manesse illuminated with roses.
It contains love songs in Middle High German

Henri Fantin-Latour was also a prolific painter of still life, particularly flowers including roses. The rose 'Fantin-Latour' was named after the artist.

Other impressionists including Claude Monet, Paul Cézanne and Pierre-Auguste Renoir have paintings of roses among their works.

Symbolism

The long cultural history of the rose has led to it being used often as a symbol. They symbolised the Houses of York and Lancaster in a conflict known as the Wars of the Roses.

Pests and Diseases

Wild roses are host plants for a number of pests and diseases. Many of these are also shared with other plants, including especially other genera of the Rosaceae.

Cultivated roses are often subject to severe damage from insect, arachnid and fungal pests and diseases. In many cases they cannot be usefully grown without regular treatment to control these problems.

Orchidaceae

The Orchidaceae are a diverse and widespread family of flowering plants, with blooms that are often colourful and fragrant, commonly known as the orchid family.

Along with the Asteraceae, they are one of the two largest families of flowering plants. The Orchidaceae have about 28,000 currently accepted species, distributed in about 763 genera. The determination of which family is larger is still under debate, because verified data on the members of such enormous families are continually in flux. Regardless, the number of orchid species nearly equals the number of bony fishes and is more than twice the number of bird species, and about four times the number of mammal species. The family also encompasses about 6–11% of all seed plants. The largest genera are *Bulbophyllum* (2,000 species), *Epidendrum* (1,500 species), *Dendrobium* (1,400 species) and *Pleurothallis* (1,000 species).

The family also includes *Vanilla* (the genus of the vanilla plant), *Orchis* (type genus), and many commonly cultivated plants such as *Phalaenopsis* and *Cattleya*. Moreover, since the introduction of tropical species into cultivation, horticulturists have produced more than 100,000 hybrids and cultivars.

Description

Orchids are easily distinguished from other plants, as they share some very evident, shared derived characteristics, or "apomorphies". Among these are: bilateral symmetry of the flower (zygomorphism), many resupinate flowers, a nearly always highly modified petal (labellum), fused stamens and carpels, and extremely small seeds.

High resolution image of orchid

Stem and Roots

Germinating seeds of the temperate orchid *Anacamptis coriophora*.
The protocorm is the first organ that will develop into true roots and leaves.

All orchids are perennial herbs that lack any permanent woody structure. They can grow according to two patterns:

- Monopodial: The stem grows from a single bud, leaves are added from the apex each year and the stem grows longer accordingly. The stem of orchids with a monopodial growth can reach several metres in length, as in *Vanda* and *Vanilla*.

- Sympodial: Sympodial orchids have a front (the newest growth) and a back (the oldest growth). The plant produces a series of adjacent shoots which grow to a certain size, bloom and then stop growing and are replaced. Sympodial orchids grow laterally rather than vertically, following the surface of their support. The growth continues by development of new leads, with their own leaves and roots, sprouting from or next to those of the previous year, as in *Cattleya*. While a new lead is developing, the rhizome may start its growth again from a so-called 'eye', an undeveloped bud, thereby branching. Sympodial orchids may have visible pseudobulbs joined by a *rhizome*, which creeps along the top or just beneath the soil.

Anacamptis lactea showing the two tubers

Terrestrial orchids may be rhizomatous or form corms or tubers. The root caps of terrestrial orchids are smooth and white.

Some sympodial terrestrial orchids, such as *Orchis* and *Ophrys*, have two subterranean tuberous roots. One is used as a food reserve for wintry periods, and provides for the development of the other one, from which visible growth develops.

In warm and constantly humid climates, many terrestrial orchids do not need pseudobulbs.

Epiphytic orchids, those that grow upon a support, have modified aerial roots that can sometimes be a few meters long. In the older parts of the roots, a modified spongy epidermis, called velamen, has the function to absorb humidity. It is made of dead cells and can have a silvery-grey, white or brown appearance. In some orchids, the velamen includes spongy and fibrous bodies near the passage cells, called tilosomes.

The cells of the root epidermis grow at a right angle to the axis of the root to allow them to get a firm grasp on their support. Nutrients for epiphytic orchids mainly come from mineral dust, organic detritus, animal droppings and other substances collecting among on their supporting surfaces.

The pseudobulb of *Prosthechea fragrans*

The base of the stem of sympodial epiphytes, or in some species essentially the entire stem, may be thickened to form a pseudobulb that contains nutrients and water for drier periods.

The pseudobulb has a smooth surface with lengthwise grooves, and can have different shapes, often conical or oblong. Its size is very variable; in some small species of *Bulbophyllum*, it is no longer than two millimeters, while in the largest orchid in the world, *Grammatophyllum speciosum* (giant orchid), it can reach three meters. Some *Dendrobium* species have long, canelike pseudobulbs with short, rounded leaves over the whole length; some other orchids have hidden or extremely small pseudobulbs, completely included inside the leaves.

With ageing, the pseudobulb sheds its leaves and becomes dormant. At this stage, it is often called a backbulb. Backbulbs still hold nutrition for the plant, but then a pseudobulb usually takes over, exploiting the last reserves accumulated in the backbulb, which eventually dies off, too. A pseudobulb typically lives for about five years. Orchids without noticeable pseudobulbs are also said to have growths, an individual component of a sympodial plant.

Leaves

Like most monocots, orchids generally have simple leaves with parallel veins, although some Vanilloideae have reticulate venation. Leaves may be ovate, lanceolate, or orbiculate, and very variable in size on the individual plant. Their characteristics are often diagnostic. They are normally alternate on the stem, often folded lengthwise along the centre ("plicate"), and have no stipules. Orchid leaves often have siliceous bodies called stegmata in the vascular bundle sheaths (not present in the Orchidoideae) and are fibrous.

The structure of the leaves corresponds to the specific habitat of the plant. Species that typically bask in sunlight, or grow on sites which can be occasionally very dry, have thick, leathery leaves and the laminae are covered by a waxy cuticle to retain their necessary water supply. Shade-loving species, on the other hand, have long, thin leaves.

The leaves of most orchids are perennial, that is, they live for several years, while others, especially those with plicate leaves as in *Catasetum*, shed them annually and develop new leaves together with new pseudobulbs.

The leaves of some orchids are considered ornamental. The leaves of the *Macodes sanderiana*, a semiterrestrial or rock-hugging ("lithophyte") orchid, show a sparkling silver and gold veining on a light green background. The cordate leaves of *Psychopsis limminghei* are light brownish-green with maroon-puce markings, created by flower pigments. The attractive mottle of the leaves of lady's slippers from tropical and subtropical Asia (*Paphiopedilum*), is caused by uneven distribution of chlorophyll. Also, *Phalaenopsis schilleriana* is a pastel pink orchid with leaves spotted dark green and light green. The jewel orchid (*Ludisia discolor*) is grown more for its colorful leaves than its white flowers.

Some orchids, as *Dendrophylax lindenii* (ghost orchid), *Aphyllorchis* and *Taeniophyllum* depend on their green roots for photosynthesis and lack normally developed leaves, as do all of the heterotrophic species.

Orchids of the genus *Corallorhiza* (coralroot orchids) lack leaves altogether and instead wrap their roots around the roots of mature trees and use specialized fungi to harvest sugars.

Vanda cultivar

Flowers

The Orchidaceae are well known for the many structural variations in their flowers.

Some orchids have single flowers, but most have a racemose inflorescence, sometimes with a large number of flowers. The flowering stem can be basal, that is, produced from the base of the tuber, like in *Cymbidium*, apical, meaning it grows from the apex of the main stem, like in *Cattleya*, or axillary, from the leaf axil, as in *Vanda*.

As an apomorphy of the clade, orchid flowers are primitively zygomorphic (bilaterally symmetrical), although in some genera like *Mormodes*, *Ludisia*, and *Macodes*, this kind of symmetry may be difficult to notice.

Dactylorhiza sambucina, Orchidoideae for reference

The orchid flower, like most flowers of monocots, has two whorls of sterile elements. The outer whorl has three sepals and the inner whorl has three petals. The sepals are usually very similar to the petals (thus called tepals, 1), but may be completely distinct.

The medial petal, called the labellum or lip (6), which is always modified and enlarged, is actually the upper medial petal; however, as the flower develops, the inferior ovary

(7) or the pedicel usually rotates 180°, so that the labellum arrives at the lower part of the flower, thus becoming suitable to form a platform for pollinators. This characteristic, called resupination, occurs primitively in the family and is considered apomorphic, a derived characteristic all Orchidaceae share. The torsion of the ovary is very evident from the longitudinal section shown (*below right*). Some orchids have secondarily lost this resupination, e.g. *Epidendrum secundum*.

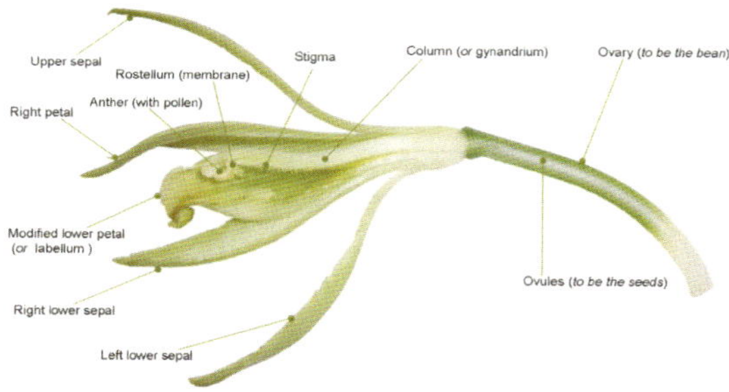

Longitudinal section of a flower of *Vanilla planifolia*

The normal form of the sepals can be found in *Cattleya*, where they form a triangle. In *Paphiopedilum* (Venus slippers), the lower two sepals are fused into a synsepal, while the lip has taken the form of a slipper. In *Masdevallia*, all the sepals are fused.

Orchid flowers with abnormal numbers of petals or lips are called peloric. Peloria is a genetic trait, but its expression is environmentally influenced and may appear random.

Laeliocattleya cultivar shows the normal form of petals.

Orchid flowers primitively had three stamens, but this situation is now limited to the genus *Neuwiedia*. *Apostasia* and the Cypripedioideae have two stamens, the central one being sterile and reduced to a staminode. All of the other orchids, the clade called *Monandria*, retain only the central stamen, the others being reduced to staminodes (4). The filaments of the stamens are always adnate (fused) to the style to form cylindrical structure called the gynostemium or column (2). In the primitive Apostasioideae, this fusion is only partial; in the Vanilloideae, it is more deep; in Orchidoideae and Epidendroideae, it is total. The stigma (9) is very asymmetrical, as all of its lobes are bent towards the centre of the flower and lie on the bottom of the column.

Pollen is released as single grains, like in most other plants, in the Apostasioideae, Cypripedioideae, and Vanilloideae. In the other subfamilies, which comprise the great majority of orchids, the anther (3) carries two pollinia.

A pollinium is a waxy mass of pollen grains held together by the glue-like alkaloid viscin, containing both cellulosic strands and mucopolysaccharides. Each pollinium is connected to a filament which can take the form of a caudicle, as in *Dactylorhiza* or *Habenaria*, or a *stipe*, as in *Vanda*. Caudicles or stipes hold the pollinia to the viscidium, a sticky pad which sticks the pollinia to the body of pollinators.

At the upper edge of the stigma of single-anthered orchids, in front of the anther cap, is the rostellum (5), a slender extension involved in the complex pollination mechanism.

As mentioned, the ovary is always inferior (located behind the flower). It is three-carpelate and one or, more rarely, three-partitioned, with parietal placentation (axile in the Apostasioideae).

In 2011, *Bulbophyllum nocturnum* was discovered to flower nocturnally.

Pollination

The complex mechanisms which orchids have evolved to achieve cross-pollination were investigated by Charles Darwin and described in *Fertilisation of Orchids* (1862). Orchids have developed highly specialized pollination systems, thus the chances of being pollinated are often scarce, so orchid flowers usually remain receptive for very long periods, rendering unpollinated flowers long-lasting in cultivation. Most orchids deliver pollen in a single mass. Each time pollination succeeds, thousands of ovules can be fertilized.

Pollinators are often visually attracted by the shape and colours of the labellum. However, some *Bulbophyllum* species attract male fruit flies (*Bactrocera* spp.) solely via a floral chemical which simultaneously acts as a floral reward (e.g. methyl eugenol, raspberry ketone, or zingerone) to perform pollination. The flowers may produce attractive odours. Although absent in most species, nectar may be produced in a spur of the labellum, or on the point of the sepals, or in the septa of the ovary, the most typical position amongst the Asparagales.

In orchids that produce pollinia, pollination happens as some variant of the following sequence: when the pollinator enters into the flower, it touches a viscidium, which promptly sticks to its body, generally on the head or abdomen. While leaving the flower, it pulls the pollinium out of the anther, as it is connected to the viscidium by the caudicle or stipe. The caudicle then bends and the pollinium is moved forwards and downwards. When the pollinator enters another flower of the same species, the pollinium has taken such position that it will stick to the stigma of the second flower, just below the rostellum, pollinating it. The possessors of orchids may be able to reproduce the process with a pencil, small paintbrush, or other similar device.

Ophrys apifera is about to self-pollinate

Some orchids mainly or totally rely on self-pollination, especially in colder regions where pollinators are particularly rare. The caudicles may dry up if the flower has not been visited by any pollinator, and the pollinia then fall directly on the stigma. Otherwise, the anther may rotate and then enter the stigma cavity of the flower (as in *Holcoglossum amesianum*).

The slipper orchid *Paphiopedilum parishii* reproduces by self-fertilization. This occurs when the anther changes from a solid to a liquid state and directly contacts the stigma surface without the aid of any pollinating agent or floral assembly.

The labellum of the Cypripedioideae is poke bonnet-shaped, and has the function of trapping visiting insects. The only exit leads to the anthers that deposit pollen on the visitor.

In some extremely specialized orchids, such as the Eurasian genus *Ophrys*, the labellum is adapted to have a colour, shape, and odour which attracts male insects via mimicry of a receptive female. Pollination happens as the insect attempts to mate with flowers.

Many neotropical orchids are pollinated by male orchid bees, which visit the flowers to gather volatile chemicals they require to synthesize pheromonal attractants. Males of such species as *Euglossa imperialis* or *Eulaema meriana* have been observed to leave their territories periodically to forage for aromatic compounds, such as cineole, to synthesize pheromone for attracting and mating with females. Each type of orchid places the pollinia on a different body part of a different species of bee, so as to enforce proper cross-pollination.

A rare achlorophyllous saprophytic orchid growing entirely underground in Australia, *Rhizanthella slateri*, is never exposed to light, and depends on ants and other terrestrial insects to pollinate it.

Catasetum, a genus discussed briefly by Darwin, actually launches its viscid pollinia with explosive force when an insect touches a seta, knocking the pollinator off the flower.

After pollination, the sepals and petals fade and wilt, but they usually remain attached to the ovary.

Asexual Reproduction

Some species, such as *Phalaenopsis*, *Dendrobium*, and *Vanda*, produce offshoots or plantlets formed from one of the nodes along the stem, through the accumulation of growth hormones at that point. These shoots are known as *keiki*.

Fruits and Seeds

Cross-sections of orchid capsules showing the longitudinal slits

The ovary typically develops into a capsule that is dehiscent by three or six longitudinal slits, while remaining closed at both ends.

The seeds are generally almost microscopic and very numerous, in some species over a million per capsule. After ripening, they blow off like dust particles or spores. They lack endosperm and must enter symbiotic relationships with various mycorrhizal basidiomyceteous fungi that provide them the necessary nutrients to germinate, so all orchid species are mycoheterotrophic during germination and reliant upon fungi to complete their lifecycles.

Closeup of a *Phalaenopsis* blossom

As the chance for a seed to meet a suitable fungus is very small, only a minute fraction of all the seeds released grow into adult plants. In cultivation, germination typically takes weeks.

Horticultural techniques have been devised for germinating orchid seeds on an artificial nutrient medium, eliminating the requirement of the fungus for germination and greatly aiding the propagation of ornamental orchids. The usual medium for the sowing of orchids in artificial conditions is agar agar gel combined with a carbohydrate energy source. The carbohydrate source can be combinations of discrete sugars or can be derived from other sources such as banana, pineapple, peach, or even tomato puree or coconut water. After the preparation of the agar agar medium, it is poured into test tubes or jars which are then autoclaved (or cooked in a pressure cooker) to sterilize the medium. After cooking, the medium begins to gel as it cools.

Distribution

Orchidaceae are cosmopolitan, occurring in almost every habitat apart from glaciers. The world's richest diversity of orchid genera and species is found in the tropics, but they are also found above the Arctic Circle, in southern Patagonia, and two species of *Nematoceras* on Macquarie Island at 54° south.

The following list gives a rough overview of their distribution:

- Oceania: 50 to 70 genera
- North America: 20 to 26 genera
- Tropical America: 212 to 250 genera
- Tropical Asia: 260 to 300 genera
- Tropical Africa: 230 to 270 genera
- Europe and temperate Asia: 40 to 60 genera

Ecology

A majority of orchids are perennial epiphytes, which grow anchored to trees or shrubs in the tropics and subtropics. Species such as *Angraecum sororium* are lithophytes, growing on rocks or very rocky soil. Other orchids (including the majority of temperate Orchidaceae) are terrestrial and can be found in habitat areas such as grasslands or forest.

Some orchids, such as *Neottia* and *Corallorhiza*, lack chlorophyll, so are unable to photosynthesise. Instead, these species obtain energy and nutrients by parasitising soil fungi through the formation of orchid mycorrhizas. The fungi involved include those that form ectomycorrhizas with trees and other woody plants, parasites such as *Armillaria*, and saprotrophs. These orchids are known as myco-heterotrophs, but were formerly (incorrectly) described as saprophytes as it was believed they gained their nutrition by breaking down organic matter. While only a few species are achlorophyllous holoparasites, all orchids are myco-heterotrophic during germination and seedling growth, and even photosynthetic adult plants may continue to obtain carbon from their mycorrhizal fungi.

Uses

As decoration in a flowerpot

A flower of a Blc. Paradise Jewel 'Flame' hybrid orchid plant. Blooms of the *Cattleya* alliance are often used in ladies' corsages.

Perfumery

The scent of orchids is frequently analysed by perfumers (using headspace technology and gas-liquid chromatography/mass spectrometry) to identify potential fragrance chemicals.

Horticulture

The other important use of orchids is their cultivation for the enjoyment of the flowers. Most cultivated orchids are tropical or subtropical, but quite a few which grow in colder climates can be found on the market. Temperate species available at nurseries include *Ophrys apifera* (bee orchid), *Gymnadenia conopsea* (fragrant orchid), *Anacamptis pyramidalis* (pyramidal orchid) and *Dactylorhiza fuchsii* (common spotted orchid).

Orchids of all types have also often been sought by collectors of both species and hybrids. Many hundreds of societies and clubs worldwide have been established. These can be small, local clubs, or larger, national organisations such as the American Orchid Society. Both serve to encourage cultivation and collection of orchids, but some go further by concentrating on conservation or research.

The term "botanical orchid" loosely denotes those small-flowered, tropical orchids belonging to several genera that do not fit into the "florist" orchid category. A few of these genera contain enormous numbers of species. Some, such as *Pleurothallis* and *Bulbophyllum*, contain approximately 1700 and 2000 species, respectively, and are often extremely vegetatively diverse. The primary use of the term is among orchid hobbyists wishing to describe unusual species they grow, though it is also used to distinguish naturally occurring orchid species from horticulturally created hybrids.

New orchids are registered with the International Orchid Register, maintained by the Royal Horticultural Society.

Use as Food

Vanilla fruits drying

The dried seed pods of one orchid genus, *Vanilla* (especially *Vanilla planifolia*), are commercially important as a flavouring in baking, for perfume manufacture and aromatherapy.

The underground tubers of terrestrial orchids [mainly *Orchis mascula* (early purple orchid)] are ground to a powder and used for cooking, such as in the hot beverage *salep* or in the Turkish frozen treat *dondurma*. The name *salep* has been claimed to come from the Arabic expression *ḥasyu al-tha`lab*, "fox testicles", but it appears more likely the name comes directly from the Arabic name *saḥlab*. The similarity in appearance to testes naturally accounts for *salep* being considered an aphrodisiac.

The dried leaves of *Jumellea fragrans* are used to flavour rum on Reunion Island.

Some saprophytic orchid species of the group *Gastrodia* produce potato-like tubers and were consumed as food by native peoples in Australia and can be successfully cultivated, notably *Gastrodia sesamoides*. Wild stands of these plants can still be found in the same areas as early aboriginal settlements, such as Ku-ring-gai Chase National Park in Australia. Aboriginal peoples located the plants in habitat by observing where bandicoots had scratched in search of the tubers after detecting the plants underground by scent.

Traditional Medicinal Uses

Orchids have been used in traditional medicine in an effort to treat many diseases and ailments. They have been used as a source of herbal remedies in China since 2800 BC. *Gastrodia elata* is one of the three orchids listed in the earliest known Chinese Materia Medica (*Shennon bencaojing*) (c. 100 AD). Theophrastus mentions orchids in his *Enquiry into Plants* (372–286 BC).

Chrysanthemum

Chrysanthemums, sometimes called mums or chrysanths, are flowering plants of the genus *Chrysanthemum* in the family Asteraceae. They are native to Asia and northeastern Europe. Most species originate from East Asia and the center of diversity is in China. There are countless horticultural varieties and cultivars.

Taxonomy

The genus once included more species, but was split several decades ago into several genera, putting the economically important florist's chrysanthemums in the genus *Dendranthema*. The naming of the genera has been contentious, but a ruling of the International Botanical Congress in 1999 changed the defining species of the genus to *Chrysanthemum indicum*, restoring the florist's chrysanthemums to the genus *Chrysanthemum*.

The other species previously included in the narrow view of the genus *Chrysanthemum* are now transferred to the genus *Glebionis*. The other genera separate from *Chrysanthemum* include *Argyranthemum*, *Leucanthemopsis*, *Leucanthemum*, *Rhodanthemum*, and *Tanacetum*.

Description

Historical painting of chrysanthemums from the *New International Encyclopedia*

Wild *Chrysanthemum* taxa are herbaceous perennial plants or subshrubs. They have alternately arranged leaves divided into leaflets with toothed or occasionally smooth edges. The compound inflorescence is an array of several flower heads, or sometimes a solitary head. The head has a base covered in layers of phyllaries. The simple row of ray florets are white, yellow or red; many horticultural specimens have been bred to bear many rows of ray florets in a great variety of colors. The disc florets of wild taxa are yellow. The fruit is a ribbed achene. Chrysanthemums, also known as 'mums', are one of the prettiest varieties of perennials that start blooming early in the autumn. This is also known as favorite flower for the month of November.

Economic uses

Ornamental uses

Chrysanthemum indicum

Different colors of *chrysanthemum x morifolium*

Example of a Japanese *bonsai* chrysanthemum

Modern cultivated chrysanthemums are showier than their wild relatives. The flower heads occur in various forms, and can be daisy-like or decorative, like pompons or buttons. This genus contains many hybrids and thousands of cultivars developed for horticultural purposes. In addition to the traditional yellow, other colors are available, such as white, purple, and red. The most important hybrid is *Chrysanthemum × morifolium* (syn. *C. × grandiflorum*), derived primarily from *C. indicum*, but also involving other species.

Over 140 cultivars of chrysanthemum have gained the Royal Horticultural Society's Award of Garden Merit (confirmed 2017).

Chrysanthemums are divided into two basic groups, garden hardy and exhibition. Garden hardy mums are new perennials capable of wintering in most northern latitudes. Exhibition varieties are not usually as sturdy. Garden hardies are defined by their ability to produce an abundance of small blooms with little if any mechanical assistance, such as staking, and withstanding wind and rain. Exhibition varieties, though, require staking, overwintering in a relatively dry, cool environment, and sometimes the addition of night lights.

The exhibition varieties can be used to create many amazing plant forms, such as large disbudded blooms, spray forms, and many artistically trained forms, such as thousand-bloom, standard (trees), fans, hanging baskets, topiary, bonsai, and cascades.

Chrysanthemum blooms are divided into 10 different bloom forms by the US National Chrysanthemum Society, Inc., which is in keeping with the international classification system. The bloom forms are defined by the way in which the ray and disk florets are arranged. Chrysanthemum blooms are composed of many individual flowers (florets), each one capable of producing a seed. The disk florets are in the center of the bloom head, and the ray florets are on the perimeter. The ray florets are considered imperfect flowers, as they only possess the female productive organs, while the disk florets are considered perfect flowers, as they possess both male and female reproductive organs.

Irregular incurves are bred to produce a giant head called an *ogiku*. The disk florets are concealed in layers of curving ray florets that hang down to create a 'skirt'. Regular incurves are similar, but usually with smaller blooms and a dense, globular form. Intermediate incurve blooms may have broader florets and a less densely flowered head.

In the reflex form, the disk florets are concealed and the ray florets reflex outwards to create a mop-like appearance. The decorative form is similar to reflex blooms, but the ray florets usually do not radiate at more than a 90° angle to the stem.

The pompon form is fully double, of small size, and very globular in form. Single and semidouble blooms have exposed disk florets and one to seven rows of ray florets.

In the anemone form, the disk florets are prominent, often raised and overshadowing the ray florets. The spoon-form disk florets are visible and the long, tubular ray florets are spatulate.

In the spider form, the disk florets are concealed, and the ray florets are tube-like with hooked or barbed ends, hanging loosely around the stem. In the brush and thistle variety, the disk florets may be visible.

In Japan a form of *bonsai* chrysanthemum was developed over the centuries. The cultivated flower has a life-span of about five years and can be kept in miniature size. Another method is to use pieces of dead wood and the flower grows over the back along the wood to give the illusion from the front that the miniature tree blooms.

Culinary uses

Yellow or white chrysanthemum flowers of the species *C. morifolium* are boiled to make a tea in some parts of Asia. The resulting beverage is known simply as chrysanthemum tea , pinyin: *júhuā chá,* in Chinese). In Korea, a rice wine flavored with chrysanthemum flowers is called *gukhwaju* .

Chrysanthemum leaves are steamed or boiled and used as greens, especially in Chinese cuisine. The flowers may be added to dishes such as *mixian* in broth, or thick snakemeat soup to enhance the aroma. Small chrysanthemums are used in Japan as a sashimi garnish.

Insecticidal uses

Pyrethrum (*Chrysanthemum* [or *Tanacetum*] *cinerariaefolium*) is economically important as a natural source of insecticide. The flowers are pulverized, and the active components, called pyrethrins, which occur in the achenes, are extracted and sold in the form of an oleoresin. This is applied as a suspension in water or oil, or as a powder. Pyrethrins attack the nervous systems of all insects, and inhibit female mosquitoes from biting. In sublethal doses they have an insect repellent effect. They are harmful to fish, but are far less toxic to mammals and birds than many synthetic insecticides. They are not persistent, being biodegradable, and also decompose easily on exposure to light. Pyrethroids such as permethrin are synthetic insecticides based on natural pyrethrum.

Persian powder is an example of industrial product of Chrysanthemum insecticide.

Environmental uses

Chrysanthemum plants have been shown to reduce indoor air pollution by the NASA Clean Air Study.

Azalea

Rhododendron 'Hinodegiri'

Azaleas are flowering shrubs in the genus *Rhododendron*, particularly the former sections *Tsutsuji* (evergreen) and *Pentanthera* (deciduous). Azaleas bloom in spring, their flowers often lasting several weeks. Shade tolerant, they prefer living near or under trees. They are part of the family Ericaceae.

Cultivation

Fifty-year-old azalea

Plant enthusiasts have selectively bred azaleas for hundreds of years. This human selection has produced over 10,000 different cultivars which are propagated by cuttings. Azalea seeds can also be collected and germinated.

Azaleas are generally slow-growing and do best in well-drained acidic soil (4.5–6.0 pH). Fertilizer needs are low; some species need regular pruning.

Azaleas are native to several continents including Asia, Europe and North America. They are planted abundantly as ornamentals in the southeastern US, southern Asia, and parts of southwest Europe.

A *George Taber* azalea

According to azalea historian Fred Galle, in the United States, *Azalea indica* (in this case, the group of plants called Southern indicas) was first introduced to the outdoor landscape in the 1830s at the rice plantation Magnolia-on-the-Ashley in Charleston, South Carolina. Magnolia's owner John Grimke Drayton imported the plants for use in his estate garden from Philadelphia, where they were grown only in greenhouses. With encouragement from Charles Sprague Sargent from Harvard's Arnold Arboretum, Magnolia Gardens was opened to the public in 1871, following the American Civil War. Magnolia is one of the oldest public gardens in America. In late March and early April, thousands visit to see the azaleas bloom in their full glory.

Disease

Azalea leafy gall can be particularly destructive to azalea leaves during the early spring. Hand picking infected leaves is the recommended method of control.

They can also be subject to phytophthora root rot in moist, hot conditions.

Cultural Significance and Symbolism

Azaleas in New Jersey

In Chinese culture, the azalea is known as "thinking of home bush" (sixiang shu) and is immortalized in the poetry of Du Fu.

The azalea is also one of the symbols of the city of São Paulo, in Brazil.

Azaleas and rhododendrons were once so infamous for their toxicity that to receive a bouquet of their flowers in a black vase was a well-known death threat.

Toxicity

In addition to being renowned for its beauty, the azalea is also highly toxic—it contains andromedotoxins in both its leaves and nectar, including honey from the nectar. Bees are deliberately fed on Azalea/Rhododendron nectar in some parts of Turkey, producing a mind-altering, potentially medicinal, and occasionally lethal honey known as "mad honey". According to the ancient Roman historian Pliny the Elder in his Natural History, an army invading Pontus in Turkey was poisoned with such honey, resulting in their defeat.

Hibiscus Rosa-sinensis

Hibiscus rosa-sinensis, known colloquially as *Chinese hibiscus, China rose, Hawaiian hibiscus*, and *shoeblackplant*, is a species of tropical hibiscus, a flowering plant in the Hibisceae tribe of the family Malvaceae, native to East Asia.

Pollen parent

Pod parent

One offspring

Another offspring

Example: two crosses of 'King Kalakua' and 'Mystic Pink'. The photographs demonstrate that the flowers of the offspring plants have almost no color characteristics of the parent plants, and few of the physical characteristics.

Nomenclature

Hibiscus rosa-sinensis was named in 1753 by Carl Linnaeus in his Species Plantarum. The Latin term *rosa-sinensis* literally means "rose of China", though it is not closely related to the true roses.

Description

Hibiscus rosa-sinensis is a bushy, evergreen shrub or small tree growing 2.5–5 m (8–16 ft) tall and 1.5–3 m (5–10 ft) wide, with glossy leaves and solitary, brilliant red flowers in summer and autumn. The 5-petaled flowers are 10 cm (4 in) in diameter, with prominent orange-tipped red anthers.

China Rose

Vegetative Characters

The root is a branched tap root. The stem is aerial, erect, green, cylindrical and branched. The leaf is simple, with alternate *phyllotaxy* and is petiolate. The leaf shape is ovate, the tip is acute and margin is serrated. *Venation* is unicostate reticulate. (Venation is branched or divergent.) Free lateral stipules are present.

Floral characters

The flower is

- Complete (bisexual),
- Actinomorphic,

- Bracteate or ebracteate,
- Bracteolate or ebracteolate,
- Pedicellate,
- Dichlamydeous,
- Regular,
- Pentamerous,
- Hypogynous - with superior Ovary, and
- Solitary.

It can bloom all year round.

Floral Formula

Br,Brl, ☿ ,K5,[C5 A∞],G5-Superior

Hibiscus flower displaying the Petals, Style, Stigma and Anthers prominently. Sepals of the Calyx (in the flower) are hidden behind the petals while they can seen nearby of a fallen flower.

Ecology

Despite its size and red hues, which are attractive to nectarivore birds, it is not visited regularly by hummingbirds when grown in the Neotropics. Generalist species, like the sapphire-spangled emerald, *Amazilia lactea*, or long-billed species, like the stripe-breasted starthroat, *Heliomaster squamosus*, are occasionally seen to visit it, however. In the subtropical and temperate Americas, hummingbirds are regularly attracted to it.

The endangered *Papilio homerus* butterfly, the largest in the western hemisphere, is known to feed on the nectar of the Hibiscus.

The stamens of the flower are partly fused into a cylinder that surrounds the style

Genetics

Hibiscus rosa-sinensis is one of many plant species with a genetic characteristic known as polyploidy, in which there are more than two complete sets of chromosomes, unlike most other species. A side effect of polyploidy is a condition where the phenotype of the offspring may be quite different from the parent, or indeed any ancestor, essentially allowing possibly random expression of all (or any) of the characteristics of all the generations that have gone before. Because of this characteristic, *H. rosa-sinensis* has become popular with hobbyists who cross and recross varieties, creating new named varieties and holding competitions to exhibit and judge the many resulting new seedlings and often strikingly unique flowers.

Uses

A white cultivar

The flowers of *Hibiscus rosa-sinensis* are edible and are used in salads in the Pacific Islands. The flower is additionally used in hair care as a preparation. It is also used to shine shoes in certain parts of India. It can also be used as a pH indicator. When used, the flower turns acidic solutions to a dark pink or magenta color and basic solutions to green. It is also used for the worship of Devi, and the red variety is especially prominent, having an important part in tantra. In Indonesia, these flowers are called "kembang sepatu", which literally means "shoe flower". In several countries the flowers are dried to use in a beverage, usually tea.

Pink Hibiscus in South India

Hibiscus rosa-sinensis is considered to have a number of medical uses in Chinese herbology. It may have some potential in cosmetic skin care; for example, an extract from the flowers of *Hibiscus rosa-sinensis* has been shown to function as an anti-solar agent by absorbing ultraviolet radiation.

Cultivation

It is widely grown as an ornamental plant throughout the tropics and subtropics. As it does not tolerate temperatures below 10 °C (50 °F), in temperate regions it is best grown under glass. However, plants in containers may be placed outside during the summer months or moved into shelter during the winter months.

Numerous varieties, cultivars, and hybrids are available, with flower colors ranging from white through yellow and orange to scarlet and shades of pink, with both single and double sets of petals. The cultivar 'Cooperi' has gained the Royal Horticultural Society's Award of Garden Merit.

Lilium

Lilium (members of which are *true lilies*) is a genus of herbaceous flowering plants growing from bulbs, all with large prominent flowers. Lilies are a group of flowering plants which are important in culture and literature in much of the world. Most species are native to the temperate northern hemisphere, though their range extends into the northern subtropics. Many other plants have "lily" in their common name but are not related to true lilies.

Description

Lilies are tall perennials ranging in height from 2–6 ft (60–180 cm). They form naked or tunicless scaly underground bulbs which are their organs of perennation. In some North American species the base of the bulb develops into rhizomes, on which numerous small bulbs are found. Some species develop stolons. Most bulbs are buried deep in the ground, but a few species form bulbs near the soil surface. Many species form stem-roots. With these, the bulb grows naturally at some depth in the soil, and each year the new stem puts out adventitious roots above the bulb as it emerges from the soil. These roots are in addition to the basal roots that develop at the base of the bulb.

Lily, petal

stamen of lilium

The flowers are large, often fragrant, and come in a wide range of colors including whites, yellows, oranges, pinks, reds and purples. Markings include spots and brush strokes. The plants are late spring- or summer-flowering. Flowers are borne in racemes or umbels at the tip of the stem, with six tepals spreading or reflexed, to give flowers varying from funnel shape to a "Turk's cap". The tepals are free from each other, and bear a nectary at the base of each flower. The ovary is 'superior', borne above the point of attachment of the anthers. The fruit is a three-celled capsule.

Seeds ripen in late summer. They exhibit varying and sometimes complex germination patterns, many adapted to cool temperate climates.

Lilium longiflorum flower – 1. Stigma, 2. Style, 3. Stamens, 4. Filament, 5. Tepal

Naturally most cool temperate species are deciduous and dormant in winter in their native environment. But a few species which distribute in hot summer and mild winter area (*Lilium candidum, Lilium catesbaei, Lilium longiflorum*) lose leaves and remain relatively short dormant in Summer or Autumn, sprout from Autumn to winter, forming dwarf stem bearing a basal rosette of leaves until, after they have received sufficient chilling, the stem begins to elongate in warming weather.

Lilium candidum seeds

The basic chromosome number is twelve (n=12).

Distribution and habitat

The range of lilies in the Old World extends across much of Europe, across most of Asia to Japan, south to India, and east to Indochina and the Philippines. In the New World they extend from southern Canada through much of the United States. They are commonly adapted to either woodland habitats, often montane, or sometimes to grassland habitats. A few can survive in marshland and epiphytes are known in tropical southeast Asia. In general they prefer moderately acidic or lime-free soils.

Ecology

Lilies are used as food plants by the larvae of some Lepidoptera species including the Dun-bar.

Cultivation

Many species are widely grown in the garden in temperate and sub-tropical regions. They may also be grown as potted plants. Numerous ornamental hybrids have been developed. They can be used in herbaceous borders, woodland and shrub plantings, and as patio plants. Some lilies, especially *Lilium longiflorum*, form important cut flower crops. These may be forced for particular markets; for instance, *Lilium longiflorum* for the Easter trade, when it may be called the Easter lily.

Lilies are usually planted as bulbs in the dormant season. They are best planted in a south-facing (northern hemisphere), slightly sloping aspect, in sun or part shade, at a depth 2½ times the height of the bulb (except *Lilium candidum* which should be planted at the surface). Most prefer a porous, loamy soil, and good drainage is essential. Most species bloom in July or August (northern hemisphere). The flowering periods of certain lily species begin in late spring, while others bloom in late summer or early autumn. They have contractile roots which pull the plant down to the correct depth, therefore it is better to plant them too shallowly than too deep. A soil pH of around 6.5 is generally safe. The soil should be well-drained, and plants must be kept watered during the growing season. Some plants have strong wiry stems, but those with heavy flower heads may need staking.

Toxicity

Some *Lilium* species are toxic to cats. This is known to be so especially for *Lilium longiflorum* though other *Lilium* and the unrelated *Hemerocallis* can also cause the same symptoms. The true mechanism of toxicity is undetermined, but it involves damage to the renal tubular epithelium (composing the substance of the kidney and secreting, collecting, and conducting urine), which can cause acute renal failure. Veterinary help should be sought, as a matter of urgency, for any cat that is suspected of eating any part of a lily – including licking pollen that may have brushed onto its coat.

Primula

A modern garden primula cultivar

Primula capitata ssp. *mooreana*

Primula denticulata

Primula farinosa

Primula vialii

Primula prolifera

Primula sieboldii

Primula × pubescens

Primula hortensis

Primula rosea

Primula veris

Primula is a genus of mainly herbaceous flowering plants in the family Primulaceae. They include the familiar wildflower of banks and verges, the primrose (*P. vulgaris*).

Other common species are *P. auricula* (auricula), *P. veris* (cowslip) and *P. elatior* (oxlip). These species and many others are valued for their ornamental flowers. They have been extensively cultivated and hybridised - in the case of the primrose, for many hundreds of years. *Primula* are native to the temperate northern hemisphere, south into tropical mountains in Ethiopia, Indonesia and New Guinea, and in temperate southern South America. Almost half of the known species are from the Himalayas.

Primula has about 500 species in traditional treatments, and more if certain related genera are included within its circumscription.

Description

Primula is a complex and varied genus, with a range of habitats from alpine slopes to boggy meadows. Plants bloom mostly during the spring, with flowers often appearing in spherical umbels on stout stems arising from basal rosettes of leaves; their flowers can be purple, yellow, red, pink, blue, or white. Some species show a white mealy bloom (farina) on various parts of the plant. Many species are adapted to alpine climates.

Ecology

Primulas are used as a food plant by the Duke of Burgundy butterfly.

Garden Hybrids and Cultivars

Primula species have been extensively cultivated and hybridised, mainly derived from *P. elatior*, *P. juliae*, *P. veris* and *P. vulgaris*. *Polyanthus* (often called *P. polyantha*) is one such group of plants, which has produced a large variety of strains in all colours, usually grown as annuals, and available as seeds or young plants.

The following hybrid varieties and cultivars have gained the Royal Horticultural Society's Award of Garden Merit:

- Crescendo Series
  - 'Crescendo Blue Shades'
  - 'Crescendo Bright Red'
  - 'Crescendo Golden'
  - 'Crescendo Pink and Rose Shades'
- Danova Series
  - 'Danova Purple, White Edge'
  - 'Danova Rose'
- 'Guinevere'
- Rainbow Series
  - 'Rainbow Blue shades'
  - 'Rainbow Cream Shades'
  - 'Rainbow Scarlet Shades'
- 'Wanda'

Delonix Regia

Delonix regia, the flame tree, is a species of flowering plant in the bean family Fabaceae, subfamily Caesalpinioideae. It is noted for its fern-like leaves and flamboyant display of flowers. In many tropical parts of the world it is grown as an ornamental tree and in English it is given the name royal poinciana or flamboyant. It is also one of several trees known as "flame tree".

This species was previously placed in the genus *Poinciana*, named for Phillippe de Longvilliers de Poincy, governor of Saint Christophe (Saint Kitts). It is a non nodulating legume.

Description

Flower, leaves & pods in Kolkata, West Bengal, India.

Delonix regia var. *flavida* is a rarer, yellow-flowered variety.

Closeup of leaves

The flowers of *Delonix regia* are large, with four spreading scarlet or orange-red petals up to 8 cm long, and a fifth upright petal called the standard, which is slightly larger and spotted with yellow and white. They appear in corymbs along and at the ends of branches. The naturally occurring variety *flavida* (Bengali: *Radhachura)* has yellow flowers. The pods are green and flaccid when young and turn dark-brown and woody. They can be up to 60 cm long and 5 cm wide. The seeds are small, weighing around 0.4 g on average. The compound leaves have a feathery appearance and are a characteristic light, bright green and are doubly pinnate. Each leaf is 30–50 cm long with 20 to 40 pairs of primary leaflets or pinnae, each divided into 10–20 pairs of secondary leaflets or pinnules.

Flower (Kibbutz Ginnosar, Israel)

Flamboyant tree (Ateneo de Manila University).

The royal poinciana (Island of Mauritius)

Distribution and Habitat

Close up of bark

Gordonvale, Queensland. Seed pods visible on upper branches.

Delonix regia is endemic to the Madagascar's dry deciduous forests but has been introduced into tropical and sub-tropical regions worldwide. In the wild it is endangered, but it is widely cultivated elsewhere.

Cultivation

The royal poinciana requires a tropical or near-tropical climate, but can tolerate drought and salty conditions. It prefers an open, free-draining sandy or loamy soil enriched with organic matter. The tree does not like heavy or clay soils and flowers more profusely when kept slightly dry.

In addition to its ornamental value, it is also a useful shade tree in tropical conditions, because it usually grows to a modest height (mostly 5 meters, but it can reach an maximum height of 12 meters) but spreads widely, and its dense foliage provides full shade. In areas with a marked dry season, it sheds its leaves during the drought, but in other areas it is virtually evergreen.

In the continental United States, it grows in South Florida, Central Florida, and in the Rio Grande Valley of South Texas. It is much loved in the Caribbean and is featured in many Dominican & Puerto Rican paintings. It can also be found in The Bahamas, Cuba, Haiti, Hawaii, Mexico (especially in the Yucatan peninsula), Nicaragua, U.S. Virgin Islands, Guam, Canary Islands, Curaçao, the Commonwealth of the Northern Mariana Islands, where it is the official tree of the islands, and Israel. It is the national flower of St. Kitts and Nevis. In Mauritius and La Réunion it announces the coming of the new year. It is very widely grown in the Northern Australia (the southern extremes previously limited to South East Queensland, although it now grows and blooms successfully in Sydney with flowering trees identified in the suburbs of Petersham, Parramatta, Guildford, Warwick Farm and Kurmond), Hong Kong, Paraguay, Peru, Spain (southern coast of Spain) the valencian coast and on all Canary Islands, Thailand, Philippines, Sri

Lanka and southern China. It is the official tree in Vietnam, Tainan, Taiwan; Xiamen, Fujian Province, People's Republic of China; and Shantou, Canton Province, People's Republic of China. National Cheng Kung University, a university located in Tainan, put royal poinciana on its emblem. It also grows throughout southern Brazil, with ornamental trees in Rio Grande do Sul (Canoas and Porto Alegre).

Royal poinciana seeds after soaking them in water for 6 days

The royal poinciana is regarded as naturalised in many of the locations where it is grown. It is a popular street tree in the suburbs of Brisbane, Australia. The tree is also found in India, where it is referred to as the May-flower tree, Gulmohar or Gul Mohr. In West Bengal (India), Odisha (India), and Bangladesh it is called Krishnachura.

The town of Peñuelas, Puerto Rico, located about 12 miles or 19 kilometers west of Ponce, is nicknamed "*El Valle de los Flamboyanes*" ("The Valley of the Poinciana Trees"), as many flamboyant trees are found along the surrounding Río Guayanes, Río Macana, and Río Tallaboa Rivers.

In Vietnam, this tree is called "Phượng vỹ", or phoenix's tail, and is a popular urban tree in much of Vietnam. Its flowering season is May–July, which coincides with the end of the school year in Vietnam. Because of this timing, the flower of poinciana is sometimes called the "flower of pupil". Hai Phong city is nicknamed "*Thành phố hoa phượng đỏ*" ("City of red poinciana").

Propagation

A bonnet macaque eating flowers

The royal poinciana is most commonly propagated by seeds. Seeds are collected, soaked in warm water for at least 24 hours, and planted in warm, moist soil in a semi-shaded, sheltered position. In lieu of soaking, the seeds can also be ‘nicked’ or ‘pinched’ (with a small scissors or nail clipper) and planted immediately. These two methods allow moisture to penetrate the tough outer casing, stimulating germination. The seedlings grow rapidly and can reach 30 cm in a few weeks under ideal conditions.

Less common, but just as effective, is propagation by semi-hardwood cuttings. Branches consisting of the current or last season’s growth can be cut into 30 cm sections and planted in a moist potting mixture. This method is slower than seed propagation (cuttings take a few months to root) but is the preferred method for ensuring new trees are true to form. As such, cuttings are a particularly common method of propagation for the rarer yellow-flowering variety of the tree.

Syringa Vulgaris

Syringa vulgaris (lilac or common lilac) is a species of flowering plant in the olive family Oleaceae, native to the Balkan Peninsula, where it grows on rocky hills. This species is widely cultivated as an ornamental and has been naturalized in other parts of Europe (including the United Kingdom, France, Germany, and Italy), as well as much of North America. It is not regarded as an aggressive species, found in the wild in widely scattered sites, usually in the vicinity of past or present human habitations.

Description

Syringa vulgaris is a large deciduous shrub or multistemmed small tree, growing to 6–7 m (20–23 ft) high, producing secondary shoots (“suckers”) from the base or roots, with stem diameters up to 20 cm (8 in), which in the course of decades may produce a small clonal thicket. The bark is grey to grey-brown, smooth on young stems, longitudinally furrowed, and flaking on older stems. The leaves are simple, 4–12 cm (2–5 in) and 3–8 cm broad, light green to glaucous, oval to cordate, with pinnate leaf venation, a mucronate apex, and an entire margin. They are arranged in opposite pairs or occasionally in whorls of three. The flowers have a tubular base to the corolla 6–10 mm long with an open four-lobed apex 5–8 mm across, usually lilac to mauve, occasionally white. They are arranged in dense, terminal panicles 8–18 cm (3–7 in) long. The fruit is a dry, smooth, brown capsule, 1–2 cm long, splitting in two to release the two-winged seeds.

Taxonomy and Naming

Syringa vulgaris was first formally described by Carl Linnaeus in 1753 and the description was published in *Species Plantarum*. The specific epithet (*vulgaris*) is a Latin word meaning “common” or “usual”.

Garden History

Lilacs—both *S. vulgaris* and *S.* × *persica* the finer, smaller "Persian lilac", now considered a natural hybrid—were introduced into northern European gardens, from Ottoman gardens, not through botanists exploring the Balkan habitats of *S. vulgaris*. The Holy Roman Emperor's ambassador, Ogier Ghiselin de Busbecq, is generally credited with supplying lilac slips to Carolus Clusius, about 1562. Well-connected botanists, such as the great herbalist John Gerard, soon had the rarity in their gardens: Gerard noted that he had lilacs growing "in very great plenty" in 1597, but lilacs were not mentioned by Shakespeare, and John Loudon was of the opinion that the Persian lilac had been introduced into English gardens by John Tradescant the elder. Tradescant's Continental source for information on the lilac, and perhaps ultimately for the plants, was Pietro Andrea Mattioli, as one can tell from a unique copy of Tradescant's plant list in his Lambeth garden, an adjunct of his *Musaeum Tradescantianum*; it was printed, though probably not published, in 1634: it lists *Lilac Matthioli*. That Tradescant's "lilac of Mattioli's" was a white one is shown by Elias Ashmole's manuscript list, *Trees found in Mrs Tredescants Ground when it came into my possession* (1662): "Syringa alba".

In the American colonies, lilacs were introduced. Peter Collinson, F.R.S., wrote to the Pennsylvania gardener and botanist John Bartram, proposing to send him some, and remarked that John Custis of Virginia had a fine "collection", which Ann Leighton interpreted as signifying common and Persian lilacs, in both purple and white, "the entire range of lilacs possible" at the time.

Cultivation

The lilac is a very popular ornamental plant in gardens and parks, because of its attractive, sweet-smelling flowers, which appear in early summer just before many of the roses and other summer flowers come into bloom.

In late summer, lilacs can be attacked by powdery mildew, specifically *Erysiphe syringae*, one of the Erysiphaceae. No fall color is seen and the seed clusters have no aesthetic appeal.

Common lilac tends to flower profusely in alternate years, a habit that can be improved by deadheading the flower clusters after the color has faded and before seeds, few of which are fertile, form. At the same time, twiggy growth on shoots that have flowered more than once or twice can be cut to a strong, outward-growing side shoot.

It is widely naturalised in western and northern Europe. In a sign of its complete naturalization in North America, it has been selected as the state flower of the state of New Hampshire, because it "is symbolic of that hardy character of the men and women of the Granite State". Additional hardiness, for Canadian gardens, was bred for in a series of *S. vulgaris* hybrids by Isabella Preston, who introduced many of the later-blooming varieties, whose later-developing flower buds are better protected from late spring frosts; the *Syringa × prestoniae* hybrids range primarily in the pink and lavender shades.

Cultivars

Most garden plants of *S. vulgaris* are cultivars, the majority of which do not exceed 4–5 m (13–16 ft) tall. Between 1876 and 1927, the nurseryman Victor Lemoine of Nancy introduced over 153 named cultivars, many of which are considered classics and still in commerce today. Lemoine's "French lilacs" extended the limited color range to include deeper, more saturated hues, and they also introduced double-flowered "sports", with the stamens replaced by extra petals.

These cultivars have gained the Royal Horticultural Society's Award of Garden Merit:

- 'Andenken an Ludwig Späth'
- 'Firmament'
- 'Katherine Havemeyer'
- 'Madame Lemoine'
- 'Vestale'

Geranium

Geranium is a genus of 422 species of flowering annual, biennial, and perennial plants that are commonly known as the cranesbills. They are found throughout the temperate regions of the world and the mountains of the tropics, but mostly in the eastern part of the Mediterranean region. The long, palmately cleft leaves are broadly circular in form. The flowers have five petals and are coloured white, pink, purple or blue, often with distinctive veining. Geraniums will grow in any soil as long as it is not waterlogged. Propagation is by semiripe cuttings in summer, by seed, or by division in autumn or spring.

The genus name is derived from the Greek 'crane'. The English name 'cranesbill' derives from the appearance of the fruit capsule of some of the species. Species in the *Geranium* genus have a distinctive mechanism for seed dispersal. This consists of a beak-like column which springs open when ripe and casts the seeds some distance. The fruit capsule consists of five cells, each containing one seed, joined to a column produced from the centre of the old flower. The common name 'cranesbill' comes from the shape of the unsprung column, which in some species is long and looks like the bill of a crane. However, many species in this genus do not have a long beak-like column.

Geraniums are eaten by the larvae of some Lepidoptera species including brown-tail, ghost moth, and mouse moth. At least several species of *Geranium* are gynodioecious.

The species *Geranium viscosissimum* (sticky geranium) is considered to be protocarnivorous.

Confusion with Pelargonium

Showing the "bill" and seed dispersal mechanism of *Geranium pratense*

Confusingly, "geranium" is also the common name of members of the genus *Pelargonium*, which are also in the Geraniaceae family and are widely grown as horticultural bedding plants. Linnaeus originally included all the species in one genus, *Geranium*, but they were later separated into two genera by Charles L'Héritier in 1789. Other former members of the genus are now classified in *Erodium*, including the plants known as filarees in North America.

The term "hardy geranium" is often applied to horticultural *Geraniums* to distinguish them from the *Pelargoniums*, which are not winter-hardy in temperate horticulture. However, not all *Geranium* species are winter-hardy.

The shape of the flowers offers one way of distinguishing between the two genera *Geranium* and *Pelargonium*. *Geranium* flowers have five very similar petals, and are thus radially symmetrical (actinomorphic), whereas *Pelargonium* (and also *Erodium*) flowers have two upper petals which are different from the three lower petals, so the flowers have a single plane of symmetry (zygomorphic).

Cultivation

A number of geranium species are cultivated for horticultural use and for pharmaceutical products. Some of the more commonly grown species include:

- *Geranium cinereum*
- *Geranium clarkei* (Clark's geranium)
- *Geranium dalmaticum*
- *Geranium endressii* (Endres's cranesbill)
- *Geranium erianthum* (wooly geranium)

- *Geranium fremontii* (Fremont's geranium)
- *Geranium himalayense*, often sold under *Geranium grandiflorum*
- *Geranium ibericum* (Caucasus geranium),
- *Geranium macrorrhizum* (bigroot cranesbill or bigroot geranium)
- *Geranium maculatum* (wild geranium)
- *Geranium maderense* (giant herb robert)
- *Geranium* × *magnificum* (showy geranium)
- *Geranium phaeum* (dusky cranesbill)
- *Geranium platypetalum* (broad-petaled geranium)
- *Geranium pratense* (meadow cranesbill)
- *Geranium psilostemon* (Armenian cranesbill)
- *Geranium renardii* (Renard geranium)
- *Geranium sanguineum* (bloody cranesbill)
- *Geranium subcaulescens* (grey cranesbill)
- *Geranium sylvaticum* (wood cranesbill)

All the above species are perennials and generally winter-hardy plants, grown for their attractive flowers and foliage. They are long-lived and most have a mounding habit, with palmately lobed foliage. Some species have spreading rhizomes. They are normally grown in part shade to full sun, in well-draining but moisture retentive soils, rich in humus. Other perennial species grown for their flowers and foliage include: *G. argenteum, G. eriostemon, G. farreri, G. nodosum, G. procurrens, G. pylzowianum, G. renardii, G. traversii, G. tuberosum, G. versicolor, G. wallichianum* and *G. wlassovianum*. Some of these are not winter-hardy in cold areas and are grown in specialized gardens like rock gardens. *Geranium* 'Johnson's Blue' is a hybrid between *G. himalayense* (southwestern China), with *G. pratense* (European meadow cranesbill).

The following hybrid cultivars have gained the Royal Horticultural Society's Award of Garden Merit.

- 'Ann Folkard'
- 'Dilys'
- 'Johnson's Blue'
- 'Mavis Simpson'
- 'Orion'
- 'Rozanne'
- 'A. T. Johnson' (*G.* × *oxonianum*)
- 'Wargrave pink' (*G.* × *oxonianum*)

References

- Eke, Tom; Al-Husainy, Sahar & Raynor, Mathew K. (2000). "The spectrum of ocular inflammation caused by Euphorbia plant sap" (PDF). Arch Ophthalmol. 118 (1): 13–16. doi:10.1001/archopht.118.1.13. PMID 10636407
- Trejo, L., M. E. Olson, P. Feria, K. M. Olsen, L. E. Eguiarte, B. Arroyo y J. A. Gruhn. 2012. "Poinsettia's wild ancestor in the Mexican dry tropics: historical, genetic, and environmental evidence." American Journal of Botany 99: 1146–1157
- Seltzer, Erica D.; Spinner, MaryAnne. "Poinsettia Facts". The Poinsettia Pages. University of Illinois Extension. Retrieved January 30, 2017
- Mark W. Chase (2001). "The origin and biogeography of Orchidaceae". In A. M. Pridgeon; P. J. Cribb; M. W. Chase; F. Rasmussen. Orchidoideae (Part 1). Genera Orchidacearum. 2. Oxford University Press. pp. 1–5. ISBN 978-0-19-850710-9
- Kimsey Lynn Siri (1980). "The behaviour of male orchid bees (Apidae, Hymenoptera, Insecta) and the question of leks". Animal Behaviour. 28 (4): 996–1004. doi:10.1016/s0003-3472(80)80088-1
- Jenny King. "The coralroot orchid". Orchids in Northern Washington State. Silvercrown Mountain Outdoor School. Retrieved 10 June 2011
- Joan Corominas (1980). Breve Diccionario Etimológico de la Lengua Castellana. Ed. Gredos. p. 328. ISBN 84-249-1332-9
- "The Legends and Traditions of Holiday Plants | Horticulture and Home Pest News". www.ipm.iastate.edu. Retrieved February 17, 2016
- Christenhusz, M. J. M. & Byng, J. W. (2016). "The number of known plants species in the world and its annual increase". Phytotaxa. Magnolia Press. 261 (3): 201–217. doi:10.11646/phytotaxa.261.3.1
- Hyam, R. & Pankhurst, R.J. (1995). Plants and their names : a concise dictionary. Oxford: Oxford University Press. ISBN 978-0-19-866189-4
- Jauron, Richard (December 8, 1995). "The Legends and Traditions of Holiday Plants". Horticulture and Home Pest News. Iowa State University. Retrieved November 27, 2017
- McGlashan, James. The Dublin University Magazine: A Literary and Political Journal. Vol. 42. July to December 1853: 286
- Zola, Nellie; Gott, Beth (1992). Koorie Plants, Koorie People: Traditional Aboriginal Food, Fibre and Healing Plants of Victoria. Koorie Heritage Trust Incorporated. p. 38. ISBN 978-1-875606-10-8
- Tom Lawrie (23 November 2010). "World's first night-flowering orchid discovered". Australian Geographic. Archived from the original on 30 November 2011. Retrieved 26 May 2013
- Hu, C. M. and G. Hao. (2011). New and noteworthy species of Primula (Primulaceae) from China. Edinburgh Journal of Botany 68(2) 297
- Brooklyn Botanic Garden (2006). Mizue Sawano: The Art of the Cherry Tree. Brooklyn Botanic Garden. p. 12. ISBN 1-889538-25-6
- Yohan Pillon & Mark W. Chase (2007). "Taxonomic exaggeration and its effects on orchid conservation". Conservation Biology. 21 (1): 263–265. doi:10.1111/j.1523-1739.2006.00573.x. PMID 17298532

5

Ornamental Trees

Many varieties of flowering trees are planted for aesthetic appeal. Some important examples are cherry blossom, magnolia, delonix regia, etc. The Japanese art form of bonsai is another significant practice that is intended for decoration of indoor and outdoor settings. This chapter closely examines the techniques and principles of practicing bonsai. It also discusses some common flowering trees that can be cultivated in floriculture.

Cherry Blossom

Cherry blossoms in Fukushima Yachounomori Garden, Tatebayashi, Gunma, Japan

A cherry blossom is the flower of any of several trees of genus *Prunus*, particularly the Japanese cherry, *Prunus serrulata*, which is called sakura after the Japanese.

Currently it is widely distributed, especially in the temperate zone of the Northern Hemisphere including Japan, Taiwan, Korea, China, West Siberia, India, Iran, and Afghanistan. Along with the chrysanthemum, the cherry blossom is considered the national flower of Japan.

Many of the varieties that have been cultivated for ornamental use do not produce fruit. Edible cherries generally come from cultivars of the related species *Prunus avium* and *Prunus cerasus*.

Cherry blossoms at the Tokyo Imperial Palace

Flower Viewing

"Hanami" is the centuries-old practice of picnicking under a blooming *sakura* or *ume* tree. The custom is said to have started during the Nara period (710–794), when it was *ume* blossoms that people admired in the beginning, but by the Heian period (794–1185) cherry blossoms came to attract more attention, and *hanami* was synonymous with *sakura*. From then on, in both waka and haiku, "flowers" meant "cherry blossoms". The custom was originally limited to the elite of the Imperial Court, but soon spread to samurai society and, by the Edo period, to the common people as well. Tokugawa Yoshimune planted areas of cherry blossom trees to encourage this. Under the *sakura* trees, people had lunch and drank sake in cheerful feasts.

Woodblock print of Mount Fuji and cherry blossom from *Thirty-six Views of Mount Fuji* by Hiroshige

Yachounomori Garden, Tatebayashi, Gunma, Japan

Every year the Japanese Meteorological Agency and the public track the *sakura zensen* (cherry blossom front) as it moves northward up the archipelago with the approach of warmer weather via nightly forecasts following the weather segment of news programs. The blossoming begins in Okinawa in January, and typically reaches Kyoto and Tokyo at the end of March or the beginning of April. It proceeds into areas at the higher altitudes and northward, arriving in Hokkaido a few weeks later. Japanese pay close attention to these forecasts and turn out in large numbers at parks, shrines and temples with family and friends to hold flower-viewing parties. *Hanami* festivals celebrate the beauty of the cherry blossom and for many are a chance to relax and enjoy the beautiful view. The custom of *hanami* dates back many centuries in Japan. The eighth-century chronicle *Nihon Shoki* records *hanami* festivals being held as early as the third century AD.

Most Japanese schools and public buildings have cherry blossom trees outside of them. Since the fiscal and school year both begin in April, in many parts of Honshu, the first day of work or school coincides with the cherry blossom season.

The Japan Cherry Blossom Association developed a list of Japan's Top 100 Cherry Blossom Spots with at least one location in every prefecture.

Varieties and Blooming

Japan has a wide variety of cherry blossoms (*sakura*); well over 200 cultivars can be found there. The following species, hybrids, and varieties are used for *sakura*:

- *Prunus apetala* var. *pilosa*
- *Prunus campanulata*
- *Prunus* × *furuseana*
- *Prunus incisa* var. *incisa*
- *Prunus incisa* var. *kinkiensis*
- *Prunus* × *introrsa*
- *Prunus* × *kanzakura*
- *Prunus* × *miyoshii*
- *Prunus padus*
- *Prunus* × *parvifolia*
- *Prunus pendula*
- *Prunus* × *sacra*
- *Prunus sargentii*
- *Prunus serrulata*
- *Prunus* × *sieboldii*
- *Prunus* × *subhirtella*
- *Prunus* × *syodoi*
- *Prunus* × *tajimensis*

- *Prunus × takenakae*
- *Prunus verecunda*
- *Prunus × yedoensis*

White cherry blossoms (sakura)

The most popular variety of cherry blossom in Japan is the *Somei Yoshino*. Its flowers are nearly pure white, tinged with the palest pink, especially near the stem. They bloom and usually fall within a week, before the leaves come out. Therefore, the trees look nearly white from top to bottom. The variety takes its name from the village of Somei (now part of Toshima in Tokyo). It was developed in the mid- to late-19th century at the end of the Edo period and the beginning of the Meiji period. The *Somei Yoshino* is so widely associated with cherry blossoms that *jidaigeki* and other works of fiction often depict the variety in the Edo period or earlier; such depictions are anachronisms.

Winter *sakura* or *fuyuzakura* (*Prunus subhirtella autumnalis*) begins to bloom in the fall and continues blooming sporadically throughout the winter. It is said to be a cross between *edohiganzakura*, the Tokyo Higan cherry (*P. incisa*) and *mamezakura* (*P. pendula*).

Other categories include *yamazakura, yaezakura,* and *shidarezakura*. The *yaezakura* have large flowers, thick with rich pink petals. The *shidarezakura*, or weeping cherry, has branches that fall like those of a weeping willow, bearing cascades of pink flowers.

Cornus Florida

Cornus florida, the flowering dogwood, is a species of flowering plant in the family Cornaceae native to eastern North America and northern Mexico. An endemic population once spanned from southernmost coastal Maine south to northern Florida and west to the Mississippi River. The tree is commonly planted as an ornamental in residential and public areas because of its showy bracts and interesting bark structure.

Classification

The flowering dogwood is usually included in the dogwood genus *Cornus* as *Cornus florida* L., although it is sometimes treated in a separate genus as *Benthamidia florida* (L.) Spach.

Two subspecies are generally recognized:

- *Cornus florida* subsp. *florida* – eastern + south-central United States.
- *Cornus florida* subsp. *urbiniana* (Rose) Rickett – northeastern Mexico (Nuevo León, Veracruz).

Description

Cornus florida inflorescence, showing four large white bracts and central flower cluster.

Flowering dogwood is a small deciduous tree growing to 10 m (33 ft) high, often wider than it is tall when mature, with a trunk diameter of up to 30 cm (1 ft). A 10-year-old tree will stand about 5 m (16 ft) tall. The leaves are opposite, simple, ovate, 6–13 cm (2.4–5.1 in) long and 4–6 cm (1.6–2.4 in) broad, with an apparently entire margin (actually very finely toothed, under a lens); they turn a rich red-brown in fall.

The flowers are individually small and inconspicuous, with four greenish-yellow bracts 4 mm (0.16 in) long. Around 20 flowers are produced in a dense, rounded, umbel-shaped inflorescence, or flower-head, 1–2 cm (0.39–0.79 in) in diameter. The flower-head is surrounded by four conspicuous large white, pink or red "petals" (actually bracts), each bract 3 cm (1.2 in) long and 2.5 cm (0.98 in) broad, rounded, and often with a distinct notch at the apex. The flowers are hermaphroditic ("perfect flowers") .

When in the wild they can typically be found at the forest edge and frequently on dry ridges. While most of the wild trees have white bracts, some selected cultivars of this tree also have pink bracts, some even almost a true red. They typically flower in early

April in the southern part of their range, to late April or early May in northern and high altitude areas. The similar Kousa dogwood (*Cornus kousa*), native to Asia, flowers about a month later.

The fruit is a cluster of two to ten separate drupes, (fused in *Cornus kousa*), each 10–15 mm (0.39–0.59 in) long and about 8 mm (0.31 in) wide, which ripen in the late summer and the early fall to a bright red, or occasionally yellow with a rosy blush. They are an important food source for dozens of species of birds, which then distribute the seeds.

Cultivation

Close up of a flower cluster showing the four pale green petals on each flower.

Flowering dogwood does best horticulturally in moist, acidic soil in a site with some afternoon shade, but good morning sun. It does not do well when exposed to intense heat sources such as adjacent parking lots or air conditioning compressors. It also has a low salinity tolerance. The hardiness zone is 5–9 and the preferred pH is between 6.0–7.0. In urban and suburban settings, care should be taken not to inflict mower damage on the trunk or roots, as this increases the tree's susceptibility to disease and pest pressure. The common flowering dogwood has been placed on the endangered species list in Ontario. In regions where dogwood anthracnose is a problem, homeowners and public land managers are encouraged to know the symptoms and inspect trees frequently. The selection of healthy, disease-free planting stock is essential and transplanting trees from the forest should be avoided. Sites should be selected for reasonably well-drained, fertile soils; full sun is recommended in high-hazard areas (such as stream or pond banks). New plantings should be mulched to a depth of 5 to 10 cm (2 to 4 in), avoiding the stem. Dead wood and leaves should be pruned and completely removed and destroyed yearly. Plants should be watered weekly during droughts, with watering done in the morning, avoiding wetting the foliage. Registered fungicides can be applied when necessary, according to manufacturers instructions and advice of local Extension Service.

Pink variety flower clusters

Flowering dogwood is grown widely throughout the temperate world.

Selected cultivars

- 'Amerika Touch-O-Pink' – large bracts, tinged pink; large leaves; good disease resistance.
- 'Appalachian Spring' – large white bracts; red fall foliage; resistant to dogwood anthracnose.
- 'Autumn Gold' – white bracts; yellow fall color.
- 'Barton' – large white bracts; blooms at early age; resistant to stem canker and powdery mildew.
- 'Bay Beauty' – double white bracts; resists heat and drought; good for Deep South.
- 'Cherokee Daybreak' – white bract; vigorous grower with variegated leaves.
- 'Cherokee Chief' – red bracts; red new growth.
- 'Cherokee Brave' – Even redder than 'Cherokee Chief', smaller bracts but dark red color; consistently resistant to powdery mildew.
- 'Cherokee Princess' – vigorous white bracts, industry standard for white flowers.
- 'Cherokee Sunset' – purplish-red bracts; variegated foliage.
- 'Gulf Coast Pink' – best pink flowering dogwood in Florida – northern part only.
- 'Hohman's Gold' – white bracts; variegated foliage.
- 'Jean's Appalachian Snow' – large, overlapping white bracts w/ green flowers; very resistant to powdery mildew.
- 'Karen's Appalachian Blush' – delicate white bracts edged in pink; some powdery mildew resistance.

- 'Kay's Appalachian Mist' – stiff, creamy white bracts; red fall foliage; good resistance to powdery mildew.
- 'Plena' – double white bracts; spot anthracnose-resistant.
- 'Purple Glory' – red bracts; purple foliage; spot anthracnose-resistant but susceptible to stem canker.
- 'Weaver White' – large white blooms; large leaves; candelabra shape; good in north-central Florida.

Propagation

Foliage during autumn

Cornus florida is easily propagated by seeds, which are sown in the fall into prepared rows of sawdust or sand, and emerge in the spring. Germination rates for good clean seed should be near 100% if seed dormancy is first overcome by cold stratification treatments for 90 to 120 days at 4 °C (39 °F). Flowering dogwood demonstrates gametophytic self-incompatibility, meaning that the plants can't self-fertilize. This is important for breeding programs as it means that it is not necessary to emasculate (remove the anthers from) *C. florida* flowers before making controlled cross-pollinations. These pollinations should be repeated every other day, as the flowers must be cross-pollinated within one or two days of opening for pollinations to be effective.

Tree in the wild in autumn.

Softwood cuttings taken in late spring or early summer from new growth can be rooted under mist if treated with 8,000 to 10,000 ppm indole-3-butyric acid (IBA). In cold climates, potted cuttings must be kept in heated cold frames or polyhouses the following winter to maintain temperatures between 0 and 7 °C (32 and 45 °F). Although rooting success can be as high as 50–85%, this technique is not commonly used by commercial growers. Rather, selected cultivars are generally propagated by T-budding in late summer or by whip grafting in the greenhouse in winter onto seedling rootstock.

Micropropagation of flowering dogwood is now used in breeding programs aiming to incorporate resistance to dogwood anthracnose and powdery mildew into horticulturally and economically important cultivars. Nodal (axillary bud) sections are established in a culture of Woody Plant Medium (WPM) amended with 4.4 µmol/L 6-Benzyladenine (BA) to promote shoot growth. Rooting of up to 83% can be obtained when 5–7 week-old microshoots are then transferred to WPM amended with 4.9 µmol/L IBA.

Magnolia Grandiflora

Magnolia grandiflora, commonly known as the southern magnolia or bull bay, is a tree of the family Magnoliaceae native to the southeastern United States, from coastal North Carolina to central Florida, and west to East Texas. Reaching 27.5 m (90 ft) in height, it is a large, striking evergreen tree, with large dark green leaves up to 20 cm (7 3/4 in) long and 12 cm (4 3/4 in) wide, and large, white, fragrant flowers up to 30 cm (12 in) in diameter.

Although endemic to the lowland subtropical forests on the Gulf and south Atlantic coastal plain, magnolia grandiflora is widely cultivated in warmer areas around the world. The timber is hard and heavy, and has been used commercially to make furniture, pallets, and veneer.

Description

Flower and foliage of *M. grandiflora*

Magnolia grandiflora is a medium to large evergreen tree which may grow 120 ft (37 m) tall. It typically has a single stem (or trunk) and a pyramidal shape.

The leaves are simple and broadly ovate, 12–20 cm ($4\frac{3}{4}$–$7\frac{3}{4}$ in) long and 6–12 cm ($2\frac{1}{4}$–$4\frac{3}{4}$ in) broad, with smooth margins. They are dark green, stiff and leathery, and often scurfy underneath with yellow-brown pubescence.

The large, showy, lemon citronella-scented flowers are white, up to 30 cm ($11\frac{3}{4}$ in) across and fragrant, with six to 12 petals with a waxy texture, emerging from the tips of twigs on mature trees in late spring.

Flowering is followed by the rose-coloured fruit, ovoid polyfollicle, 7.5–10 cm (3–$3\frac{7}{8}$ in) long, and 3–5 cm ($1\frac{1}{4}$–2 in) wide.

Exceptionally large trees have been reported in the far southern United States. The national champion is a specimen in Smith County, Mississippi, that stands an incredible 37 m (121 ft). Another record includes a 35-m-high specimen from the Chickasawhay District, De Soto National Forest, in Mississippi, which measured 17.75 ft in circumference at breast height, from 1961, and a 30-m-tall tree from Baton Rouge, which reached 18 ft in circumference at breast height.

Taxonomy

Magnolia grandiflora was one of the many species first described by Linnaeus in the 10th edition of his *Systema Naturae* in 1759, basing his description on the earlier notes of Miller. He did not select a type specimen. Its specific epithet is derived from the Latin words *grandis* "big", and *flor-* "flower".

M. grandiflora is most commonly known as southern magnolia, a name derived from its range in the Southern United States. Many broadleaved evergreen trees are known as bays for their resemblance to the leaves of the red bay (*Persea borbonia*), with this species known as the bull bay for its huge size or alternatively because cattle have been reported eating its leaves. Laurel magnolia, evergreen magnolia, large-flower magnolia or big laurel are alternative names. The timber is known simply as magnolia.

M. grandiflora fruit

Distribution and Habitat

Southern magnolias are native to the Southeastern United States, from coastal North Carolina south to central Florida, and then west to East Texas. It is found on the edges of bodies of water and swamps, in association with sweetgum (*Liquidambar styraciflua*), water oak (*Quercus nigra*), and black tupelo (*Nyssa sylvatica*). In more sheltered habitats, it grows as a large tree, but can be a low shrub when found on coastal dunes. It is killed by summer fires, and is missing from habitats that undergo regular burning.

In Florida, it is found in a number of different ecological areas that are typically shady and have well-draining soils; it is also found in hummocks, along ravines, on slopes, and in wooded floodplains. Despite preferring sites with increased moisture, it does not tolerate inundation. It grows on sand-hills in maritime forests, where it is found growing with live oaks and saw palmetto. In the eastern United States, it has become an escape, and has become naturalized in the tidewater area of Virginia, the Washington, D.C. metropolitan area and locally in other areas outside of its historically natural range.

Ecology

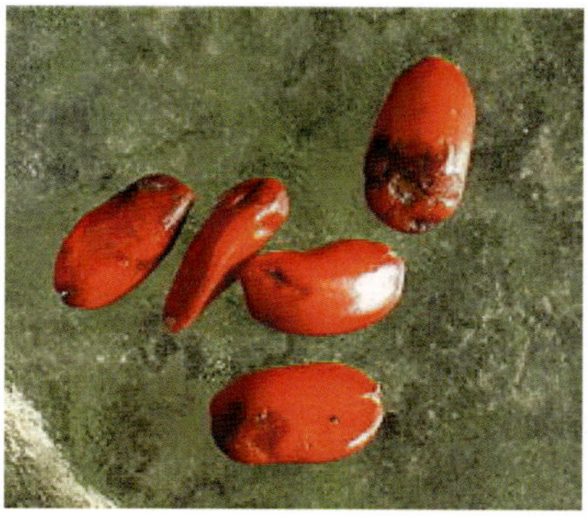

Individual seeds

Magnolia grandiflora can produce seed by 10 years of age, although peak seed production is achieved closer to 25 years of age. Around 50% of seeds can germinate, and they are spread by birds and mammals. Squirrels, opossums, quail, and turkey are known to eat the seeds.

Cultivation and uses

The plant collector Mark Catesby, the first in North America, brought *M. grandiflora* to Britain in 1726, where it entered cultivation and overshadowed *M. virginiana*, which had been collected a few years earlier. It had also come to France, the French having collected it in the vicinity of the Mississippi River in Louisiana. It was glowingly described by Philip Miller in his 1731 work *The Gardeners' Dictionary*. One of the earliest people to cultivate it in Europe was Sir John Colliton of Exeter in Devon; scaffolding and tubs surrounded his tree, where gardeners propagated its branches by layering, the daughter plants initially selling for five guineas each (but later falling to half a guinea).

Tree planted 1807 at Jardin des plantes in Nantes

Catesby's *Laurel tree of Carolina* by G.D. Ehret

It is often planted in university campuses and allowed to grow into a large tree, either with dependent branches, or with the lower branches removed to display the bare trunks. It is also espaliered against walls, which improves its frost-hardiness.

Lagerstroemia

Lagerstroemia commonly known as crape myrtle or crepe myrtle, is a genus of around 50 species of deciduous and evergreen trees and shrubs native to the Indian subcontinent, southeast Asia, northern Australia, and parts of Oceania, cultivated in warmer climates around the world. It is a member of the family Lythraceae, which are also known as the loosestrife family. The genus is named after the Swedish merchant Magnus von Lagerström, a director of the Swedish East India Company who supplied Carl Linnaeus with plants he collected. These flowering trees are beautifully colored and are often planted both privately and commercially as ornamentals.

Description

Crepe myrtles are chiefly known for their colorful and long-lasting flowers which occur in summer. Most species of *Lagerstroemia* have sinewy, fluted stems and branches with a mottled appearance that arises from having bark that sheds throughout the year. The leaves are opposite and simple, with entire margins, and vary from 5–20 cm (2–8 in). While all species are woody in nature, they can range in height from over 30 m (100 ft) to under 30 cm (1 ft); most, however, are small to medium multiple-trunked trees and shrubs. The leaves of temperate species provide autumn color.

A 4 m (12 ft) crape myrtle in Lutherville, Maryland

Red crepe myrtle -- *Lagerstroemia*

Flowers are borne in summer and autumn in panicles of crinkled flowers with a crepe-like texture. Colors vary from deep purple to red to white, with almost every shade in between. Although no blue-flowered varieties exist, the flowers trend toward the blue end of the spectrum with no orange or yellow except in stamens and pistils. The fruit is a capsule, green and succulent at first, then ripening to dark brown or black dryness. It splits along six or seven lines, producing teeth much like those of the calyx, and releases numerous, small, winged seeds.

In their respective climates, both subtropical and tropical species are common in domestic and commercial landscapes. The timber of some species has been used to manufacture bridges, furniture, and railway sleepers, but in Vietnam's Cat Tien National Park, the dominant stands of *Lagerstroemia calyculata* in secondary forest are thought to have survived (after episodes of logging) due to the low quality of wood. *Lagerstroemia* species are used as food plants by the larvae of some Lepidoptera (moth and butterfly) species including *Endoclita malabaricus*.

The leaves of *L. parviflora* are fed on by the *Antheraea paphia* moth which produces the tassar silk, a form of wild silk of commercial importance in India.

L. microcarpa leaves

Laburnum

Laburnum, sometimes called golden chain or golden rain, is a genus of two species of small trees in the subfamily Faboideae of the pea family Fabaceae. The species are *Laburnum anagyroides*—common laburnum and *Laburnum alpinum*—alpine laburnum. They are native to the mountains of southern Europe from France to the Balkans.

Some botanists include a third species, *Laburnum caramanicum*, but this native of southeast Europe and Anatolia is usually treated in a distinct genus *Podocytisus*, more closely allied to the *Genisteae* (brooms).

Description

The *Laburnum* trees are deciduous. The leaves are trifoliate, somewhat like a clover; the leaflets are typically 2–3 cm (1–1 in) long in *L. anagyroides* and 4–5 cm (1.5–2 in) long in *L. alpinum*.

They have yellow pea-flowers in pendulous leafless racemes 10–40 cm (4–15.5 in) long in spring, which makes them very popular garden trees. In *L. anagyroides*, the racemes are 10–20 cm (4–8 in) long, with densely packed flowers; in *L. alpinum* the racemes are 20–30 cm (8–12 in) long, but with the flowers sparsely along the raceme. The fruit develops as a pod and is extremely poisonous.

The yellow flowers are responsible for the old poetic name 'golden chain tree' (also spelled golden chaintree or goldenchain tree).

Laburnum tree in full flower

All parts of the plant are poisonous, although mortality is very rare. Symptoms of laburnum poisoning may include intense sleepiness, vomiting, convulsive movements, coma, slight frothing at the mouth and unequally dilated pupils. In some cases, diarrhea is very severe, and at times the convulsions are markedly tetanic. The main toxin in the plant is cytisine, a nicotinic receptor agonist.

It is used as a food plant by the larvae of some Lepidoptera species, including the buff-tip.

Cultivation

Laburnum species and hybrids are cultivated as ornamental trees for gardens and parks. They are also trained as espaliers on pergolas, for ceilings of pendant flowers in season. In its natural form, *Laburnum* is a shrubby, multi-branched tree, but it is often pruned to maintain a single trunk which displays the smooth green bark.

Gardeners are advised to remove the spent seedpods after flowering because they sap the strength of the tree and are the most poisonous part. Generally *Laburnum* does not perform well in hot climates, and has a reduced life-span if grown in climates with winters warmer than Zone 7 (0 °F/-17 °C minimum). Afternoon shade and the occasional deep watering are advisable in areas with hot, dry summers. They do best in climates with moderate winter and summer temperatures, ideally Oceanic climates like those of the Pacific Northwest and Northern Europe. *Laburnum* trees are ubiquitous in England, Wales, Scotland and Ireland, where they are commonly planted as lawn specimens or in shrub borders.

Most garden specimens are of the hybrid between the two species, *Laburnum ×watereri* 'Vossii' (Voss's laburnum), which combines the longer racemes of *L. alpinum* with the denser flowers of *L. anagyroides*; it also has the benefit of low seed production. It has gained the Royal Horticultural Society's Award of Garden Merit.

Delonix Regia

Delonix regia, the flame tree, is a species of flowering plant in the bean family Fabaceae, subfamily Caesalpinioideae. It is noted for its fern-like leaves and flamboyant display of flowers. In many tropical parts of the world it is grown as an ornamental tree and in English it is given the name royal poinciana or flamboyant. It is also one of several trees known as "flame tree".

This species was previously placed in the genus *Poinciana*, named for Phillippe de Longvilliers de Poincy, the governor of Saint Christophe (Saint Kitts). It is a non nodulating legume.

Description

The flowers of *Delonix regia* are large, with four spreading scarlet or orange-red petals up to 8 cm long, and a fifth upright petal called the standard, which is slightly larger and spotted with yellow and white. They appear in corymbs along and at the ends of branches. The naturally occurring variety *flavida* (Bengali: *Radhachura)* has yellow flowers. The pods are green and flaccid when young and turn dark-brown and woody. They can be up to 60 cm long and 5 cm wide. The seeds are small, weighing around 0.4 g on average. The compound (doubly pinnate) leaves have a feathery appearance and are a characteristic light, bright green. Each leaf is 30–50 cm long with 20 to 40 pairs of primary leaflets or pinnae, each divided into 10–20 pairs of secondary leaflets or pinnules.

Distribution and Habitat

Delonix regia is endemic to the Madagascar's dry deciduous forests but has been introduced into tropical and sub-tropical regions worldwide. In the wild it is endangered, but it is widely cultivated elsewhere.

Cultivation

The royal poinciana requires a tropical or near-tropical climate, but can tolerate drought and salty conditions. It prefers an open, free-draining sandy or loamy soil enriched with organic matter. The tree does not like heavy or clay soils and flowers more profusely when kept slightly dry.

(Delonix regia)

Close-up of part of a leaf

Flower, leaves & pods (Kolkata, West Bengal, India)

Delonix regia var. *flavida* is a rarer, yellow-flowered variety.

Flower (Kibbutz Ginnosar, Israel)

Flowering tree (Island of Mauritius)

In addition to its ornamental value, it is also a useful shade tree in tropical conditions, because it usually grows to a modest height (mostly 5 meters, but it can reach an maximum height of 12 meters) but spreads widely, and its dense foliage provides full shade. In areas with a marked dry season, it sheds its leaves during the drought, but in other areas it is virtually evergreen.

Close up of bark

Seed pods visible on upper branches (Gordonvale, Queensland, Australia)

In the continental United States, it grows in South Florida, Central Florida, and in the Rio Grande Valley of South Texas. It is much loved in the Caribbean and is featured in many Dominican & Puerto Rican paintings. It can also be found in Belize,The Bahamas, Cuba, Haiti, Hawaii, Mexico (especially in the Yucatan peninsula), Nicaragua, U.S. Virgin Islands, Guam, Canary Islands, Curaçao, the Commonwealth of the Northern Mariana Islands, where it is the official tree of the islands, and Israel. It is the national flower of St. Kitts and Nevis. In Mauritius and La Réunion it announces the coming of the new year. It is very widely grown in the Northern Australia (the southern extremes previously limited to South East Queensland, although it now grows and blooms successfully in Sydney with flowering trees identified in the suburbs of Petersham, Parramatta, Guildford, Warwick Farm and Kurmond), Hong Kong, Paraguay, Peru, Spain (southern coast of Spain) the valencian coast and on all

Canary Islands, Thailand, Philippines, Sri Lanka and southern China. It is the official tree in Vietnam, Tainan, Taiwan; Xiamen, Fujian Province, People's Republic of China; and Shantou, Canton Province, People's Republic of China. National Cheng Kung University, a university located in Tainan, put royal poinciana on its emblem. It also grows throughout southern Brazil, with ornamental trees in Rio Grande do Sul (Canoas and Porto Alegre).

Seeds after soaking them in water for 6 days

The royal poinciana is regarded as naturalised in many of the locations where it is grown. It is a popular street tree in the suburbs of Brisbane, Australia. The tree is also found in India, where it is referred to as the May-flower tree, Gulmohar or Gul Mohr. In West Bengal (India), Odisha (India), and Bangladesh it is called Krishnachura.

The town of Peñuelas, Puerto Rico, located about 12 miles or 19 kilometers west of Ponce, is nicknamed "*El Valle de los Flamboyanes*" ("The Valley of the Poinciana Trees"), as many flamboyant trees are found along the surrounding Río Guayanes, Río Macana, and Río Tallaboa Rivers.

In Vietnam, this tree is called "Phượng vỹ", or phoenix's tail, and is a popular urban tree in much of Vietnam. Its flowering season is May–July, which coincides with the end of the school year in Vietnam. Because of this timing, the flower of poinciana is sometimes called the "flower of pupil". Hai Phong city is nicknamed "*Thành phố hoa phượng đỏ*" ("City of red poinciana")

Propagation

The royal poinciana is most commonly propagated by seeds. Seeds are collected, soaked in warm water for at least 24 hours, and planted in warm, moist soil in a semi-shaded, sheltered position. In lieu of soaking, the seeds can also be 'nicked' or 'pinched' (with a small scissors or nail clipper) and planted immediately. These two methods allow moisture to penetrate the tough outer casing, stimulating germination. The seedlings grow rapidly and can reach 30 cm in a few weeks under ideal conditions.

A bonnet macaque eating flowers

Less common, but just as effective, is propagation by semi-hardwood cuttings. Branches consisting of the current or last season's growth can be cut into 30 cm sections and planted in a moist potting mixture. This method is slower than seed propagation (cuttings take a few months to root) but is the preferred method for ensuring new trees are true to form. As such, cuttings are a particularly common method of propagation for the rarer yellow-flowering variety of the tree.

Bonsai

Bonsai is a Japanese art form using cultivation techniques to produce small trees in containers that mimic the shape and scale of full size trees. Similar practices exist in other cultures, including the Chinese tradition of *penzai* or *penjing*, and the miniature living landscapes of Vietnamese *Hòn Non Bộ*. The Japanese tradition dates back over a thousand years.

Japanese white pine from the National Bonsai & Penjing Museum at the United States National Arboretum.

Bonsai at the Omiya Bonsai Art Museum

Bonsai at the National Bonsai & Penjing Museum at the United States National Arboretum

The purposes of bonsai are primarily contemplation for the viewer, and the pleasant exercise of effort and ingenuity for the grower. By contrast with other plant cultivation practices, bonsai is not intended for production of food or for medicine. Instead, bonsai practice focuses on long-term cultivation and shaping of one or more small trees growing in a container.

A bonsai is created beginning with a specimen of source material. This may be a cutting, seedling, or small tree of a species suitable for bonsai development. Bonsai can be created from nearly any perennial woody-stemmed tree or shrub species that produces true branches and can be cultivated to remain small through pot confinement with crown and root pruning. Some species are popular as bonsai material because they have characteristics, such as small leaves or needles, that make them appropriate for the compact visual scope of bonsai.

The source specimen is shaped to be relatively small and to meet the aesthetic standards of bonsai. When the candidate bonsai nears its planned final size it is planted in a display pot, usually one designed for bonsai display in one of a few accepted shapes and proportions. From that point forward, its growth is restricted by the pot environment. Throughout the year, the bonsai is shaped to limit growth, redistribute foliar vigor to areas requiring further development, and meet the artist's detailed design.

The practice of bonsai is sometimes confused with dwarfing, but dwarfing generally refers to research, discovery, or creation of plant cultivars—especially coniferous subjects such as juniper and spruce—that are permanent, genetic miniatures of existing species. Bonsai does not require genetically dwarfed trees, but rather depends on growing small trees from regular stock and seeds. Bonsai uses cultivation techniques like pruning, root reduction, potting, defoliation, and grafting to produce small trees that mimic the shape and style of mature, full-size trees.

Display

A bonsai display presents one or more bonsai specimens in a way that allows a viewer to see all the important features of the bonsai from the most advantageous position. That position emphasizes the bonsai's defined "front", which is designed into all bonsai. It places the bonsai at a height that allows the viewer to imagine the bonsai as a full-size tree seen from a distance, siting the bonsai neither so low that the viewer appears to be hovering in the sky above it nor so high that the viewer appears to be looking up at the tree from beneath the ground. Noted bonsai writer Peter Adams recommends that bonsai be shown as if "in an art gallery: at the right height; in isolation; against a plain background, devoid of all redundancies such as labels and vulgar little accessories."

Bonsai displayed on an outdoor bench - note automated watering apparatus

A Seiju elm bonsai on display with a shitakusa of miniature hosta and a hanging scroll.

For outdoor displays, there are few aesthetic rules. Many outdoor displays are semi-permanent, with the bonsai trees in place for weeks or months at a time. To avoid damaging the trees, therefore, an outdoor display must not impede the amount of sunlight needed for the trees on display, must support watering, and may also have to block excessive wind or precipitation. As a result of these practical constraints, outdoor displays are often rustic in style, with simple wood or stone components. A common design is the bench, sometimes with sections at different heights to suit different sizes of bonsai, along which bonsai are placed in a line. Where space allows, outdoor bonsai specimens are spaced far enough apart that the viewer can concentrate on one at a time. When the trees are too close to each other, aesthetic discord between adjacent trees of different sizes or styles can confuse the viewer, a problem addressed by exhibition displays.

Exhibition displays allow many bonsai to be displayed in a temporary exhibition format, typically indoors, as would be seen in a bonsai design competition. To allow many trees to be located close together, exhibition displays often use a sequence of small alcoves, each containing one pot and its bonsai contents. The walls or dividers between the alcoves make it easier to view only one bonsai at a time. The back of the alcove is a neutral color and pattern to avoid distracting the viewer's eye. The bonsai pot is almost always placed on a formal stand, of a size and design selected to complement the bonsai and its pot.

Indoors, a formal bonsai display is arranged to represent a landscape, and traditionally consists of the featured bonsai tree in an appropriate pot atop a wooden stand, along with a shitakusa (companion plant) representing the foreground, and a hanging scroll representing the background. These three elements are chosen to complement each other and evoke a particular season, and are composed asymmetrically to mimic nature. When displayed inside a traditional Japanese home, a formal bonsai display will often be placed within the home's tokonoma or formal display alcove. An indoor display is usually very temporary, lasting a day or two, as most bonsai are intolerant of indoor conditions and lose vigor rapidly within the house.

Containers

Assorted bonsai pots

A variety of informal containers may house the bonsai during its development, and even trees that have been formally planted in a bonsai pot may be returned to growing boxes from time to time. A large growing box can house several bonsai and provide a great volume of soil per tree to encourage root growth. A training box will have a single specimen, and a smaller volume of soil that helps condition the bonsai to the eventual size and shape of the formal bonsai container. There are no aesthetic guidelines for these development containers, and they may be of any material, size, and shape that suit the grower.

Completed trees are grown in formal bonsai containers. These containers are usually ceramic pots, which come in a variety of shapes and colors and may be glazed or unglazed. Unlike many common plant containers, bonsai pots have drainage holes in the bottom surface to complement fast-draining bonsai soil, allowing excess water to escape the pot. Growers cover the holes with a screening to prevent soil from falling out and to hinder pests from entering the pots from below. Pots usually have vertical sides, so that the tree's root mass can easily be removed for inspection, pruning, and replanting, although this is a practical consideration and other container shapes are acceptable.

There are alternatives to the conventional ceramic pot. Multi-tree bonsai may be created atop a fairly flat slab of rock, with the soil mounded above the rock surface and the trees planted within the raised soil. In recent times, bonsai creators have also begun to fabricate rock-like slabs from raw materials including concrete and glass-reinforced plastic. Such constructed surfaces can be made much lighter than solid rock, can include depressions or pockets for additional soil, and can be designed for drainage of water, all characteristics difficult to achieve with solid rock slabs. Other unconventional containers can also be used, but in formal bonsai display and competitions in Japan, the ceramic bonsai pot is the most common container.

For bonsai being shown formally in their completed state, pot shape, color, and size are chosen to complement the tree as a picture frame is chosen to complement a painting. In general, containers with straight sides and sharp corners are used for formally shaped plants, while oval or round containers are used for plants with informal designs. Many aesthetic guidelines affect the selection of pot finish and color. For example, evergreen bonsai are often placed in unglazed pots, while deciduous trees usually appear in glazed pots. Pots are also distinguished by their size. The overall design of the bonsai tree, the thickness of its trunk, and its height are considered when determining the size of a suitable pot.

Some pots are highly collectible, like ancient Chinese or Japanese pots made in regions with experienced pot makers such as Tokoname, Japan or Yixing, China. Today many potters worldwide produce pots for bonsai.

Size Classifications

Japanese bonsai exhibitions and catalogs frequently refer to the size of individual bonsai specimens by assigning them to size classes (see table below). Not all sources agree on the exact sizes or names for these size ranges, but the concept of the ranges is well-established and useful to both the cultivation and the aesthetic understanding of the trees. A photograph of a bonsai may not give the viewer an accurate impression of the tree's real size, so printed documents may complement a photograph by naming the bonsai's size class. The size class implies the height and weight of the tree in its container.

In the very largest size ranges, a recognized Japanese practice is to name the trees "two-handed", "four-handed", and so on, based on the number of men required to move the tree and pot. These trees will have dozens of branches and can closely simulate a full-size tree. The very largest size, called "imperial", is named after the enormous potted trees of Japan's Imperial Palace.

At the other end of the size spectrum, there are a number of specific techniques and styles associated solely with the smallest common sizes, *mame* and *shito*. These techniques take advantage of the bonsai's minute dimensions and compensate for the limited number of branches and leaves that can appear on a tree this small.

Common names for bonsai size classes		
Large bonsai		
Common name	**Size class**	**Tree Height**
Imperial bonsai	Eight-handed	152–203 cm (60–80 in)
Hachi-uye	Six-handed	102–152 cm (40–60 in)
Dai	Four-handed	76–122 cm (30–48 in)
Omono	Four-handed	76–122 cm (30–48 in)
Medium-size bonsai		
Common name	**Size class**	**Tree Height**
Chiu	Two-handed	41–91 cm (16–36 in)
Chumono	Two-handed	41–91 cm (16–36 in)
Katade-mochi	One-handed	25–46 cm (10–18 in)
Miniature bonsai		
Common name	**Size class**	**Tree Height**
Komono	One-handed	15–25 cm (6–10 in)
Shohin	One-handed	13–20 cm (5–8 in)
Mame	Palm size	5–15 cm (2–6 in)
Shito	Fingertip size	5–10 cm (2–4 in)
Keshitsubo	Poppy-seed size	3–8 cm (1–3 in)

Indoor Bonsai

Carissa macrocarpa developed by Frank Okamura

Indoor bonsai are bonsai which are cultivated for the indoor environment. Traditionally, bonsai are temperate climate trees grown outdoors in containers. Kept in the artificial environment of a home, these trees weaken and die. But a number of tropical and sub-tropical tree species will survive and grow indoors. Some of these are suited to bonsai aesthetics and can be shaped much as traditional outdoor bonsai are.

Note that bonsai and similar practices like *penjing*, *hòn non bộ*, and *saikei* all involve the long-term cultivation of small trees in containers. The term *bonsai* is generally used in English as an umbrella term for all miniature trees in containers or pots. In this article *bonsai* should be understood to include any container-grown tree that is raised indoors and regularly styled or shaped, not just one being maintained in the Japanese bonsai tradition.

Indoor vs. traditional Bonsai

The largest difference between indoor and traditional bonsai is, of course, the enjoyment of an attractive, fully leaved plant in winter instead of a dormant, leafless tree. Other differences include the faster growth rate of tropical plants, which accelerate all steps of the bonsai evolution. Moss covering, a common soil covering for outdoor bonsai, will not survive indoor conditions.

Bonsai vs. other Forms of House Plant

Compared to the usual potted house plant, bonsai are rooted in a much smaller amount of soil. Consequently, they require more frequent watering and feeding. This form is therefore best suited for drought-resistant species. Compared to usual house gardening, bonsais require a lot more pruning, both of branches and roots. This often requires a significant shift in attitude for house gardeners.

The spiritual benefits of bonsai cultivation (*bonsai no kokoro*) are available equally to classical and indoor bonsai gardeners

Cultivation and Care

Many of the techniques used to grow and shape outdoor bonsai apply equally to indoor bonsai. For example, shaping techniques like pruning, wiring, and grafting work much the same for both types of bonsai. But indoor bonsai also have some special requirements for cultivation and care.

Christmas Cactus (*Schlumbergera*)

Lighting

An indoor room comfortably lit for human use provides too little light for most species of tree to grow. Few species will thrive with less than 500 lux available several hours per day. Successful indoor bonsai cultivation requires either selecting from the short list of low-light-tolerant trees, or providing additional lighting for the trees. Insufficient light may not kill some species outright, but will make their growth so slow that the bonsai shaping techniques cannot be used: the plant will not grow back after pruning or leaf trimming.

The simplest way to provide extra light is to place the bonsai close to an outside window. Care must be taken not to harm the plant either with cold air entering through the window, or with high radiated heat from direct exposure to the sun. More controllable light, and more flexibility in situating the bonsai, can be obtained using artificial lighting. Fluorescent lighting (preferably with bulbs radiating growth-friendly spectra) and light-emitting diode lighting can give sufficient light to support a significant number of indoor bonsai species. These lights also have the advantage of blending with other indoor lighting, so that the plants can be grown in normal living quarters. For plants needing the highest amount of light, or for large numbers of bonsai being maintained indoors, a space dedicated to bonsai cultivation can support high-intensity lamps and optionally the special ventilation or cooling often needed for their use.

Temperature

Acer buergerianum, U.S. National Bonzai and Penjing Museum

Suitable temperatures for indoor bonsai extend from standard room temperature downward. The best temperature range for a tree is determined by its species and, less directly, by its normal range in the wild. Tropical trees are usually tolerant of room temperature year-round but cannot tolerate temperatures approaching freezing, as might occur near an open window in cold weather. Semi-tropical and Mediterranean-climate trees often grow better when temperatures drop well below room temperature during winter months. The lower temperature, combined with shorter periods of daylight, triggers an expected annual dormancy which many non-tropical species need in order to thrive. These non-tropical trees need a cool location in the winter, such as a cool window ledge or "cold room" area in the house. In addition to the need for seasonal temperature variation, many non-tropicals grow better when there is a distinct difference between day temperatures (warm) and night (cooler).

Humidity

Indoor conditions, particularly in homes outside the tropics, imply very low humidity. Both air conditioning and room heating reduce air humidity significantly. Some tree species, such as ficus sp. with their waxy leaves, are tolerant of low humidity. Many species, however, require additional humidification to survive indoors. As with lighting, some humidification solutions can be integrated into the regular living area and some require a dedicated space. Local humidity can be raised by locating the bonsai containers over a tray of water, or by placing them in a bed of dampened sand. For better effect, the bonsai can be placed in enclosures, such as terrariums, that are humidified from within. For the maximum humidity, a dedicated room with high-capacity evaporative or misting humidifiers can support the most sensitive bonsai.

Plants Suitable for Indoor Bonsai

Tropical

The creation of bonsai is limited only by the imagination and talent of the gardener, although some species are much more suitable than others. Members of the genus *Ficus* are among the most versatile, while many succulents can be grown in a similar fashion. Here is an incomplete list of the most popular species.

Crassula ovata

- *Crassula ovata*: The jade plant is a very robust and drought-resistant house plant. The miniature cultivars like the baby jade plant (*C. ovata arborescens*) is considered the best plant for a first bonsai. This plant will sprout on old wood. Thus, an old specimen can be pruned back to a stockier shape with thick trunk. It is kept dry in winter, placed outdoors in summer for full growth. Its roots are thin and cannot be exposed.

- *Cycas revoluta*: This plant is often used as an indoor bonsai. Though it does not have a traditional tree shape, it can give an image of a tropical island.
- *Dracaena marginata*: The dragon plant has an interesting palm-like shape. It can sprout on old wood. It does not tolerate root exposure.
- *Ficus benjamina*: The Weeping Fig is a popular indoor tree that lends itself to the classical, upright form. It is one of the few tropicals that are accepted as "true" bonsai. The miniature cultivars like 'Too Little' are also well suited for bonsai. It forms aerial roots and can be shaped as a banyan tree. Ficus are intolerant to branch down-pruning; one must start with a small tree and keep it small. They are sensitive to stress.
- *Ficus neriifolia*: According to Jerry Meislik, "the most useful fig for bonsai is the willow leafed fig. The small leaf is in excellent scale for bonsai and the tree has good branch ramification, good basal rootage and excellent aerial root formation."
- *Portulacaria afra*: The dwarf jade looks a lot like a baby jade plant and is used similarly.

Schefflera arboricola

- *Schefflera arboricola*: The Hawaiian umbrella tree is a popular, hardy houseplant that is ideal for irregular, banyan or roots-on-rock forms. Since it can sprout on old wood, an old specimen can be pruned back to a stockier shape with thick trunk and roots. It tolerates root exposure very well, is drought-resistant and requires a moderate amount of light. Under high humidity conditions, it produces aerial roots and can therefore be shaped as a banyan tree.
- *Schlumbergera*: The Christmas cactus does not have a real trunk but easily lends itself to a cascade-type bonsai shape. It tolerates shade, not drought.
- *Tamarindus indica* The tamarind has recently become popular in bonsai culture, frequently used in Asian countries like Indonesia, Taiwan and the Philippines. In the last Japan Airlines World Bonsai competition, Mr. Budi Sulistyo of Indonesia won the second prize with an ancient tamarind bonsai.

Small succulents may be used as accent plants:

- Rhipsalis (Hatiora) salicornioides

Other Climatic Origins

With proper care, a number of non-tropical plants can also thrive as indoor bonsai.

Camellia japonica, U.S. National Bonsai and Penjing Museum

- *Araucaria heterophylla*: The Norfolk Pine is a needle-bearing evergreen. It does well indoors and is often seen in commercial buildings, but is not easily shaped to bonsai styles.
- *Ardisia crenata*: The spear-flower will flower and produce berry-like fruit when grown indoors.
- *Bougainvillea*: This flowering shrub, native to South America, will grow indoors with sufficient light.
- *Bucida molineti*: The black olive is native to southern Florida in the USA, and has a compact growth habit suited to bonsai forms.
- *Boxwood*: Found on nearly every continent, boxwood varieties are tough but attractive bonsai. The varieties *Buxus harlandii* and *Buxus microphylla* (Kingsville boxwood) have tiny leaves and develop mature-looking bark, supporting bonsai aesthetics.
- *Calliandra haematocephala* (Red powderpuff): The *Calliandra* family includes evergreen bushes growing in tropical and subtropical areas worldwide. The *haematocephala* species has small needle-like leaves and unusual-looking feathery blossom that appear in spring and summer.
- *Camellia japonica* (Common camellia), *Camellia sinensis*: In the mountains of Japan and Korea, trees of the *Camellia* family grow up to 10 to 12 meters high. *Camellia sinensis* in particular suit bonsai treatment, with relatively small leaves and flowers.

- *Carmona microphylla*: The sturdy trunk and glossy green leaves of the Fukien Tea make an attractive indoor bonsai. Tiny white flowers appear regularly, and a characteristically rough or gnarled bark gives an appearance of age on even young bonsai.
- *Casuarina equisetifolia* (Horsetail tree): The horsetail tree grows up to 25 meters in the wild, but can be cultivated indoors as a compact bonsai. Its foliage is needle-like although the plant is considered a deciduous tree, and it can be shaped to resemble a spruce or pine.
- *Citrus japonica 'Hindsii'* (Kumquat): This fruiting tree is one of the kumquat varieties, and relatives with small leaves suitable for indoor bonsai include *Citrus japonica 'Margarita'*, *Citrus japonica 'Japonica'* (Marumi kumquat), × *Citrofortunella mitis*, *Citrus aurantifolia* (Key lime), and *Citrofortunella microcarpa* (Calamondin). The trees flower and may even bear fruit in pot cultivation.
- *Cupressus macrocarpa*: The Monterey Cypress is one of the few evergreen conifers that can be grown indoors. Trunk thickness is difficult to develop, but the fine juniper-like foliage suits even small bonsai.
- *Feroniella lucida*: Native to southeast Asia, this tree flowers and produces edible fruit. It is a slow grower in a container, but small leaves and a rough trunk make it very suitable for bonsai.
- *Fraxinus uhdei* (Tropical ash): Though the leaf clusters do not match common bonsai aesthetics, this tree can be shaped into an effective bonsai, particularly as the trunk ages and develops character.
- *Ligustrum sp.*: Some privet species, including *Ligustrum japonica* and *Ligustrum rotundifolium*, are suitable subjects for indoor bonsai.

Serissa japonica, U.S. National Banzai and Penjing Museum

- *Serissa*: This is a delicate flowering plant with tiny leaves that can grow indoors year round.

- *Syzygium paniculatum*, syn. *Eugenia paniculata*: A flowering and fruiting evergreen tree with slender leaves, this native of Australia grows to 6 meters in the wild. As a bonsai, it can be shaped in most styles suiting deciduous trees.
- *Ulmus parvifolia*: The Chinese elm is often used as outdoor bonsai, and it can be raised indoors as well.

Bonsai Aesthetics

A bald cypress in the formal upright style.

Bonsai aesthetics are the aesthetic goals and characteristics of the Japanese tradition in the art of growing a miniature tree in a container. Many Japanese cultural characteristics, particularly the influence of Zen Buddhism and the expression of *wabi* or *sabi*, inform the bonsai tradition in that culture. As well, a lengthy catalog of conventional tree shapes and styles helps provide cohesion to the Japanese styling tradition. A number of other cultures around the globe have adopted the Japanese approach to bonsai, and while some variations have begun to appear, most hew closely to the rules and design philosophies of the Japanese tradition.

A Japanese Black Pine in an informal style.

The aesthetics of *penjing*, a Chinese form of container-grown tree, are distinct from those of bonsai and are discussed elsewhere. The aesthetics of *saikei*, Japanese multi-tree landscapes in a container, are also distinct and are not described in this article.

John Naka's famous bonsai *Goshin*, showing some deadwood effects.

A Blue Atlas Cedar (Cedrus libani var. atlantica) bonsai on display at the National Bonsai & Penjing Museum at the United States National Arboretum.

Over centuries of practice, the Japanese bonsai aesthetic has encoded some important methods and aesthetic guidelines. Like the type of aesthetic rules that govern, for example, Western common practice period music, bonsai's guidelines help practitioners work within an established tradition with some assurance of success. Guidelines alone do not guarantee a successful result. Nevertheless, these design rules can rarely be broken without reducing the impact of the bonsai specimen.

Bonsai styles

A key design practice in bonsai is a set of commonly understood, named styles that describe canonical tree and setting designs. These well-known styles provide a convenient shorthand means for communicating about existing bonsai and for designing new ones. Bonsai styles describe a number of basic attributes of a bonsai, such as the angle and straightness of its trunk, its branch configuration, and the number of trees in the bonsai container.

The system of styles serves many purposes, some practical, some aesthetic. In their simplest and most common application, styles provide a form of shorthand description for bonsai specimens. Predefined styles also aid the designer in making a development plan for a pre-bonsai tree. The untrained specimen may have characteristics that suggest or rule out certain styles. The designer can evaluate the pre-bonsai specimen against the catalog of accepted styles to determine what branches to remove or reshape, what foliage to remove or encourage, and what detailed shaping to apply to trunk and branches.

As with all aesthetic rules or guidelines, the various accepted styles will guide a bonsai designer, but are not completely deterministic. The species of the bonsai, the age of the tree when it began bonsai training, the tree's pre-existing shape and structure, even the bonsai artist's training and preferences, strongly affect the shape of the resulting bonsai. These competing influences ensure that the style system acts mostly as a creative aid, not a dominating constraint, in producing a finished bonsai.

General Aesthetic Principles

The main aim of bonsai aesthetic practices is to create miniature trees with an air of age in their overall shapes, proportions, and details. The quintessential bonsai is a single, dwarfed tree in a small container. It has the appearance of a mature tree, but not of a completely natural one. Instead, a designer or artist has manipulated the shape and surfaces of the tree to enhance or exaggerate the tree's apparent age, and also to give it a defined "front" from which it is meant to be viewed. Anyone questioning the effect of the bonsai designer's work can test the quality of the design by viewing it from the rear, where exactly the same trunk and branches will generally look awkward, cluttered, or otherwise unattractive.

No trace of the Artist

At the same time, the designer's touch must not be apparent to the viewer. If a branch is removed in shaping the tree, the scar will be placed at the "back" of the tree where it cannot be seen. Alternatively, the tree will not be shown until the scar has been covered by years of bark growing over it, or a stub of the branch will remain to be cleaned and shaped into looking like it was broken by wind or lightning. Similarly, wiring should be removed or at least concealed when the bonsai is shown, and must leave no permanent marks on the branch or bark.

Visual Balance

Other guidelines address the balance of visual weight among the trunk, roots, foliage, and branches. The extensive catalog of recognized tree styles form part of this set of guidelines. The term "balance" here may refer to either:

- static visual balance, where careful application of symmetry leads to a stable and restful shape (like the formal upright, or *Chokkan*, style), or
- dynamic visual balance, which may arise from an asymmetric shape or one that implies instability and movement (like the cascade, or *Kengai*, style).

The trunk, roots, foliage, and branches are manipulated through a variety of techniques to meet the designer's goals of visual balance. Negative spaces (the "empty space" between solid elements like branches or foliage) are also shaped and proportioned to appear in balance. In almost all designs, the viewer can see completely through the tree's negative spaces to the background behind it. In this combination of positive and negative shapes, bonsai aesthetics overlap to a certain extent with the aesthetics of sculpture.

Proportion Among Elements

Another general guideline touches on the proportion of the bonsai's various elements. The most prized proportions mimic those of a full-grown tree as closely as possible. Slender branches with heavy leaves or needles that are out of proportion are avoided, as is a thin trunk with thick branches. One of the few exceptions to this guideline is that flowers and fruit (on trees that produce them) are not considered to be flawed if they appear too large for the tree.

Flexibility of the Rules

One or more of the accepted rules of bonsai form can be bent or broken for a particular tree without destroying its fundamental aesthetic and artistic impact. In fact, going beyond the prescribed rules allows aesthetic growth in the bonsai art, as seen in many of the masterpieces created by Masahiko Kimura and Kunio Kobayashi.

General Aesthetic Guidelines

The following characteristics are desirable in many Japanese bonsai and other styles of container-grown trees, whatever the style:

Gravitas

This is the trait which all of the remaining points of aesthetics seek to create. It is a sense of physical weight, the illusion of mass, the appearance of maturity or advanced age, and the elusive quality of dignity. Many of the formal rules of bonsai help the grower create a tree that expresses *wabi* or *sabi*, or portrays an aspect of *mono no aware.*

Miniaturization

By definition, a bonsai is a tree which is kept small enough to be container-grown while otherwise fostered to have a mature appearance. Bonsai can be classified according to size. *Mame* are ideally less than 10 cm (4 inches) tall and can be held in the palm of the hand. *Shohin* are about 25 cm (10 inches) tall, while other bonsai are larger and can not be easily moved. For both practical and aesthetic reasons, the guidelines outlined here are generally most effective and most often applied to larger bonsai, while the smallest specimens of bonsai may adhere to no rules other than "miniature tree" and "grown in a container".

Lignification

This refers to enhancing the "woody-ness" of a bonsai's trunk and branches so that they have a mature appearance. This typically means the bark surface is encouraged to become rough and dark-colored. In some cases this aesthetic technique will vary, as in a birch tree bonsai attaining the white colour and exfoliating bark of a mature specimen.

Asymmetry

Bonsai aesthetics discourage strict symmetry in branch and root placement. Radial symmetry is nearly always broken by the requirement for a clear "front", which exposes the tree's trunk and major branches. The left, right, and back sides will have more branches than the front. Left-right (bilateral) symmetry across the trunk is also discouraged, and designers work to alternate branches among the left, right, and back parts of the tree without ever placing two branches at the same height or extending two branches the same distance away from the trunk.

Leaf Reduction

Leaf reduction is related to the general miniaturization described above but is something which varies over the life cycle of a particular bonsai. For example, a bonsai's leaves might be allowed to attain full size for many years in order to encourage vigor and growth of trunk, roots, and branches. It is usually desirable to attain a degree of leaf reduction prior to exhibiting a bonsai. Leaf reduction may be encouraged by pruning and is sometimes achieved by the total defoliation of a bonsai during one part of its growing season. Conifer needles are more difficult to reduce than other sorts of foliage.

Nebari

Also known as "buttressing", *nebari* is the visible spread of roots above the growing medium at the base of a bonsai. Nebari help a bonsai seem grounded and well-anchored and make it look mature, akin to a full-sized tree.

Ramification

Ramification is the splitting of branches and twigs into smaller ones. It is encouraged by pruning and may be integrated with practices that promote leaf reduction.

Deadwood

Bonsai artists sometimes create or emphasize the appearance of dead wood on a bonsai tree, reflecting the occasional presence of dead branches or snags on full-sized trees. Two specific styles of deadwood are *jin* and *shari*. The presence of deadwood is not as common as most of the other points mentioned here, but can be used very effectively on selected tree species and bonsai styles. See deadwood techniques for more details.

Curvature

Trunk and branch curvature or contortion is an optional goal. Bonsai can achieve a sense of age while remaining straight and upright, but many bonsai rely upon curvature of the trunk to build the illusion of weight and age. Curvature of the trunk that occurs between the roots and the lowest branch is known as *tachiagari*. Branches are also curved and recurved to help them fit the designer's requirement for "positive space", and to separate small branches so that they do not cross or collide.

Concrete Aesthetic Guidelines

To support the general goals and principles of bonsai aesthetics, a number of detailed heuristics are taught in the bonsai tradition and documented in its literature (somewhat similar to the bonsai styles). Example guidelines include:

- The tree will appear in a formal container, relatively small compared to the tree.
- Except for the tree(s) and optional patches of moss, no other vegetation should appear in a bonsai container.
- Except for the soil, allowed vegetation, and optionally natural-looking rocks, no other object should appear in a bonsai container.
- The tree will have a distinct "front" from which it is intended to be viewed.
- The trunk should taper significantly from base to top.
- The tree's rootage should be exposed at the base of the trunk and should flare wider than the trunk as it enters the ground.
- No visible root should cross another.
- Branches should begin about one-third of the way up the trunk, and be continuous from there to the tip of the trunk (this guideline is specifically broken for the literati, or *Bunjin-gi*, style).
- Branch size should diminish from the base to the top of the tree.
- No major tree branch should cross the trunk when viewed from the tree's "front".
- Branch ramification, particularly in deciduous trees, should increase towards the tip of each branch.
- Branch shape should reflect the weight of age, particularly in conifers, and branches may be shaped to tend downwards toward the tip in support of this practice.
- The trunk may be a straight vertical shape or may be contorted in different directions over its length, but in styles where the tip of the tree is above the container, the tip should tilt slightly forward at the top (toward the viewer).

- Foliage (leaves or needles) should be small and to scale with the tree and its branches.
- All trees in a multi-tree bonsai planting should be of the same species.

Bonsai Styles

Bonsai is a Japanese art form using miniature trees grown in containers. Similar practices exist in other cultures, including the Chinese tradition of *penjing* from which the art originated, and the miniature living landscapes of Vietnamese *hòn non bộ*, but this article describes the Japanese tradition.

The Japanese art of bonsai dates back over a thousand years, and has evolved its own unique aesthetics and terminology. A key design practice in bonsai is a set of commonly understood, named styles that describe canonical tree and setting designs. These well-known styles provide a convenient shorthand means for communicating about existing bonsai and for designing new ones.

Concept of Styles

Cascade style conifer

Forest style Black Hills Spruce

Twin-trunk style smoothleaf elm (Ulmus minor)

Informal upright style Juniper

Slant style Hinoki cypress

Driftwood style Sierra juniper (Juniperus occidentalis var. australis)

Styles can be grouped based on different criteria, such as the trunk orientation or the number of trunks in the bonsai specimen. Some of the major style groupings include:

- Trunk orientation. A frequently used set of styles describe the orientation of the bonsai tree's main trunk. Different terms are used for a tree with its apex directly over the center of the trunk's entry into the soil (these are the upright styles, including *chokkan* and *moyogi*), slightly to the side of that center (e.g., *sho-shakan*), deeply inclined to one side (e.g., *chu-shakan* and *dai-shakan*), and inclined below the point at which the trunk of the bonsai enters the soil (the cascade or *kengai* styles).
- Trunk and bark surface. A number of styles describe the trunk shape and bark finish. For example, a bonsai with a twisted trunk is *nebikan* (also *nejikan*), and one with a vertical split or hollows is *sabakan*. The deadwood bonsai styles identify trees with prominent dead branches or trunk scarring.
- Trunk and root placement. Although most bonsai trees are planted directly into the soil, there are styles describing trees planted on rock. For example, the root-over rock style is *deshojo,* and the style in which trees are rooted wholly within (atop or on the sides of) a large rock is *ishizuki*.
- Multiple trunks. While the majority of bonsai specimens feature a single tree, there are well-established style categories for specimens with multiple trunks. Within these styles, a bonsai can be classified by number of trunks alone (e.g., *sokan* for a double trunk from a single root, *soju* for two separate trees, *sambon-yose* for three trees, and so on). The configuration of the trunks can also be described by specific styles, including raft (*ikadabuji* or *ikadabuki*) and sinuous (*netsunagari*) styles for multiple trees growing from a connected root, and the general term *yose-ue* for multiple unconnected trees in large number.

These terms are not mutually exclusive, and a single bonsai specimen can exhibit more than one style characteristic. When a bonsai specimen falls into multiple style categories, the common practice is to describe it by the dominant or most striking characteristic. For example, an informal upright tree with prominent areas of missing bark and trunk scarring will be described as a *sharimiki* rather than a *moyogi*.

Purposes

The system of styles serves many purposes, some practical, some aesthetic. In their simplest and most common application, styles provide a form of shorthand description for bonsai specimens. The brief style term appears in catalog descriptions, usually with a species identifier, and thereby compactly describes the subject bonsai. Style names can also be used to group comparable specimens in bonsai viewing and competition. Even considering the styles simply as descriptive labels, the system still simplifies bonsai teaching and learning, and provides widely understood terms for public communications about bonsai.

Predefined styles also aid the designer in making a development plan for a pre-bonsai tree. The untrained specimen may have characteristics that suggest or rule out certain styles. For example, a crooked trunk makes a tree unsuitable for the formal upright style, and suggests to the designer that the tree may be trained better as an informal upright or a slanted style instead. A damaged or highly asymmetrical tree may not appear suitable for bonsai development, yet may be adapted to an uncommon style like windswept or raft, which both work for trees that have branches only on one side of the trunk. Some tree species are not suitable for some styles: a bonsai artist working with a deciduous tree will not produce bonsai in the cascade style, for example. The designer can evaluate the pre-bonsai specimen against the catalog of accepted styles to determine what branches to remove or reshape, what foliage to remove or encourage, and what detailed shaping to apply to trunk and branches.

Although the styles will guide a bonsai designer, they are not completely deterministic. A review of actual bonsai from competition catalogs will reveal that even highly regarded specimens rarely meet every rule laid out for their style. The species of the bonsai, the age of the tree when it began bonsai training, the tree's pre-existing shape and structure, even the bonsai artist's training and preferences, strongly affect the shape of the resulting bonsai. These competing influences ensure that the style system acts mostly as a creative aid, not a dominating constraint, in producing a finished bonsai.

Bonsai Cultivation and Care

A Trident Maple bonsai from the National Bonsai & Penjing Museum at the United States National Arboretum.

Bonsai cultivation and care involves the long-term cultivation of small trees in containers, called *bonsai* in the Japanese tradition of this art form. Similar practices exist in other Japanese art forms and in other cultures, including *saikei* (Japanese), *penjing* (Chinese), and *hòn non bộ* (Vietnamese). Trees are difficult to cultivate in containers, which restrict root growth, nutrition uptake, and resources for transpiration (primarily soil moisture). In addition to the root constraints of containers, bonsai trunks, branches, and foliage are extensively shaped and manipulated to meet aesthetic goals.

Specialized tools and techniques are used to protect the health and vigor of the subject tree. Over time, the artistic manipulation of small trees in containers has led to a number of cultivation and care approaches that successfully meet the practical and the artistic requirements of bonsai and similar traditions.

The term *bonsai* is generally used in English as an umbrella term for all miniature trees in containers or pots. In this article *bonsai* should be understood to include any container-grown tree that is regularly styled or shaped, not just one being maintained in the Japanese bonsai tradition.

Bonsai can be created from nearly any perennial woody-stemmed tree or shrub species which produces true branches and remains small through pot confinement with crown and root pruning. Some species are popular as bonsai material because they have characteristics, such as small leaves or needles, that make them appropriate for the compact visual scope of bonsai. Bonsai cultivation techniques are different from other tree cultivation techniques in allowing mature (though miniature) trees to grow in small containers, to survive with extremely restricted root and canopy structures, and to support comprehensive, repeated styling manipulations.

Sources of Bonsai Material

All bonsai start with a specimen of source material, a plant that the grower wishes to train into bonsai form. Bonsai practice is an unusual form of plant cultivation in that growth from seeds is rarely used to obtain source material. To display the characteristic aged appearance of a bonsai within a reasonable time, the source plant is often partially grown or mature stock. A specimen may be selected specifically for bonsai aesthetic characteristics it already possesses, such as great natural age for a specimen collected in the wild, or a tapered, scar-free trunk from a nursery specimen. Alternatively, it may be selected for non-aesthetic reasons, such as known hardiness for the grower's local climate or low cost (as in the case of collected materials).

Propagation

Plant cuttings can be rooted and grown as potential bonsai.

While any form of plant propagation could generate bonsai material, a few techniques are favored because they can quickly produce a relatively mature trunk with well-placed branches.

Cuttings. In taking a cutting, part of a growing plant is cut off and placed in a growing medium to develop roots. If the part that is cut off is fairly thick, like a mature branch, it can be grown into an aged-looking bonsai more quickly than can a seed. Unfortunately, thinner and younger cuttings tend to strike roots more easily than thicker or more mature ones. In bonsai propagation, cuttings usually provide source material to be grown for some time before training.

Layering. Layering is a technique in which rooting is encouraged from part of a plant, usually a branch, while it is still attached to the parent plant. After rooting, the branch is removed from the parent and grown as an independent entity. For bonsai, both ground layering and air layering can create a potential bonsai, by transforming a mature branch into the trunk of a new tree. The point at which rooting is encouraged can be close to the location of side branches, so the resulting rooted tree can immediately have a thick trunk and low branches, characteristics that complement bonsai aesthetics.

Commercial Bonsai Growers

Commercial bonsai growers may use any of the other means of obtaining starter bonsai material, from seed propagation to collecting expeditions, but they generally sell mature specimens that display bonsai aesthetic qualities already. The grower trains the source specimens to a greater or lesser extent before sale, and the trees may be ready for display as soon as they are bought. Those who purchase commercially grown bonsai face some challenges, however, particularly of buying from another country. If the purchaser's local climate does not closely match the climate in which the bonsai was created, the plant will have difficulties surviving and thriving. As well, importing living plant material from a foreign source is often closely controlled by import regulations and may require a license or other special import arrangement on the buyer's part. If a local commercial bonsai grower does not exist, buying from a distant one may be unsatisfactory.

Nursery Stock

A plant nursery is an agricultural operation where (non-bonsai) plants are propagated and grown to usable size. Nursery stock may be available directly from the nursery, or may be sold in a garden centre or similar resale establishment. Nursery stock is usually young but fully viable, and is often potted with sufficient soil to allow plants to survive a season or two before being transplanted into a more permanent location. Because the nursery tree is already pot-conditioned, it can be worked on as a bonsai immediately. The large number of plants that can be viewed in a single visit to a nursery or garden centre allows the buyer to identify plants with better-than-average bonsai characteristics. According to Peter Adams, a nursery visit "offers the opportunity to choose an instant trunk". One issue with nursery stock is that many specimens are shaped into popular forms, such as the standard or half-standard forms, with several feet of clear trunk rising from the roots. Without branches low on the trunk, it is difficult for a source specimen to be trained as bonsai.

Collecting

Collecting bonsai consists of finding suitable bonsai material in its original wild situation, successfully moving it, and replanting it in a container for development as bonsai. Collecting may involve wild materials from naturally treed areas, or cultivated specimens found growing in yards and gardens. For example, mature landscape plants being discarded from a building site can provide excellent material for bonsai. Hedgerow trees, grown for many years but continually trimmed to hedge height, provide heavy, gnarled trunks for bonsai collectors. In locations close to a tree line (the line beyond which trees do not grow, whether due to altitude, temperature, soil moisture, or other conditions), aged and naturally dwarfed survivors can be found.

The main benefit of collecting bonsai specimens is that collected materials can be mature, and will display the natural marks and forms of age, which makes them more suitable for bonsai development than the young plants obtained through nurseries. Low cost is another potential benefit, with a tree harvest license often being more economical than purchase of nursery trees. Some of the difficulties of collecting include finding suitable specimens, getting permission to remove them, and the challenges of keeping a mature tree alive while transplanting it to a bonsai pot.

Styling Techniques

This juniper makes extensive use of both jin (deadwood branches) and shari (trunk deadwood).

Bonsai are carefully styled to maintain miniaturization, to suggest age, and to meet the artist's aesthetic goals. Tree styling also occurs in a larger scale in other practices like topiary and niwaki. In bonsai, however, the artist has close control over every feature of the tree, because it is small and (in its container) easily moved and worked on. The greater scale of full-sized trees means that styling them may be restricted to pruning and shaping the exterior volume once per growing season, never pruning within the canopy nor bending and forming individual branches. In contrast, in a bonsai being prepared for display, each leaf or needle may be subject to decision regarding pruning or retention, and every branch and twig may be formed and wired into place each year. Given these differences in scope and purpose, bonsai styling uses a number of styling techniques either unique to bonsai or (if used in other forms of plant cultivation) applied in ways particularly suitable to meet the goals of bonsai development.

Leaf Trimming

This technique involves selective removal of leaves (for most varieties of deciduous tree) or needles (for coniferous trees and some others) from a bonsai's branches. A common aesthetic technique in bonsai design is to expose the tree's branches below groups of leaves or needles (sometimes called "pads") by removing downward-growing material. In many species, particularly coniferous ones, this means that leaves or needles projecting below their branches must be trimmed off. For some coniferous varieties, such as spruce, branches carry needles from the trunk to the tip and many of these needles may be trimmed to expose the branch shape and bark. Needle and bud trimming can also be used in coniferous trees to force back-budding on old wood, which may not occur naturally in many conifers. Along with pruning, leaf trimming is the most common activity used for bonsai development and maintenance, and the one that occurs most frequently during the year.

Pruning

The small size of the tree and some dwarfing of foliage result from pruning the trunk, branches, and roots. Pruning is often the first step in transforming a collected plant specimen into a candidate for bonsai. The top part of the trunk may be removed to make the tree more compact. Major and minor branches that conflict with the designer's plan will be removed completely, and others may be shortened to fit within the planned design. Pruning later in the bonsai's life is generally less severe, and may be done for purposes like increasing branch ramification or encouraging growth of non-pruned branches. Although pruning is an important and common bonsai practice, it must be done with care, as improper pruning can weaken or kill trees. Careful pruning throughout the tree's life is necessary, however, to maintain a bonsai's basic design, which can otherwise disappear behind the uncontrolled natural growth of branches and leaves.

Wiring

Wrapping copper or aluminium wire around branches and trunks allows the bonsai designer to create the desired general form and make detailed branch and leaf placements.

When wire is used on new branches or shoots, it holds the branches in place until they lignify (convert into wood). The time required is usually 6–9 months or one growing season for deciduous trees, but can be several years for conifers like pines and spruce, which maintain their branch flexibility through multiple growing seasons. Wires are also used to connect a branch to another object (e.g., another branch, the pot itself) so that tightening the wire applies force to the branch. Some species do not lignify strongly, and some specimens' branches are too stiff or brittle to be bent easily. These cases are not conducive to wiring, and shaping them is accomplished primarily through pruning.

Extensive wiring can be seen on this bonsai specimen.

Clamping

For larger specimens, or species with stiffer wood, bonsai artists also use mechanical devices for shaping trunks and branches. The most common are screw-based clamps, which can straighten or bend a part of the bonsai using much greater force than wiring can supply. To prevent damage to the tree, the clamps are tightened a little at a time and make their changes over a period of months or years.

Grafting

In this technique, new growing material (typically a bud, branch, or root) is introduced to a prepared area under the bark of the tree. There are two major purposes for grafting in bonsai. First, a number of favorite species do not thrive as bonsai on their natural root stock and their trunks are often grafted onto hardier root stock. Examples include Japanese red maple and Japanese black pine. Second, grafting allows the bonsai artist to add branches (and sometimes roots) where they are needed to improve or complete a bonsai design. There are many applicable grafting techniques, none unique to bonsai, including branch grafting, bud grafting, thread grafting, and others.

Defoliation

Short-term dwarfing of foliage can be accomplished in certain deciduous bonsai by partial or total defoliation of the plant partway through the growing season. Not all species can survive this technique. In defoliating a healthy tree of a suitable species, most or all of the leaves are removed by clipping partway along each leaf's petiole (the thin stem that connects a leaf to its branch). Petioles later dry up and drop off or are manually removed once dry. The tree responds by producing a fresh crop of leaves. The new leaves are generally much smaller than those from the first crop, sometimes as small as half the length and width. If the bonsai is shown at this time, the smaller leaves contribute greatly to the bonsai aesthetic of dwarfing. This change in leaf size is usually not permanent, and the leaves of the following spring will often be the normal size. Defoliation weakens the tree and should not be performed in two consecutive years.

Deadwood

Bonsai growers create or shape dead wood using techniques such as *jin* and *shari* to simulate age and maturity in a bonsai. Jin is the term used when the bark from an entire branch is removed to create the impression of a snag of deadwood. Shari denotes stripping bark from areas of the trunk to simulate natural scarring from a broken limb or lightning strike. In addition to stripping bark, deadwood techniques may also involve the use of tools to scar the deadwood or to raise its grain, and the application of chemicals (usually lime sulfur) to bleach and preserve the exposed deadwood.

Care

Small trees grown in containers, like bonsai, require specialized care. Unlike most houseplants, flowering shrubs, and other subjects of container gardening, tree species in the wild generally grow individual roots up to several meters long and root structures encompassing hundreds or thousands of liters of soil. In contrast, a typical bonsai container allows a fraction of a meter for root extension, and holds 2 to 10 liters of soil and root mass. Branch and leaf (or needle) growth in trees is also large-scale in nature. Wild trees typically grow 5 meters or taller when mature, while the largest bonsai rarely exceed 1 meter and most specimens are significantly smaller. These size differences affect maturation, transpiration, nutrition, pest resistance, and many other aspects of tree biology. Maintaining the long-term health of a tree in a container requires a number of specialized care techniques.

Growing Environment

Most bonsai species are trees and shrubs that must by nature grow outdoors. They require temperature, humidity, and sunlight conditions approximating their native climate year round. The skill of the grower can help bonsai from outside the local hardiness zone survive and even thrive, but doing so takes careful watering, shielding of selected bonsai from excessive sunlight or wind, and possibly protection from winter conditions (e.g., through the use of cold frames or winter greenhouses).

Common bonsai species (particularly those from the Japanese tradition) are temperate climate trees from hardiness zones 7 to 9, and require moderate temperatures, moderate humidity, and full sun in summer with a dormancy period in winter that may need be near freezing. They do not thrive indoors, where the light is generally too dim, and humidity often too low, for them to grow properly. Only during their dormant period can they safely be brought indoors, and even then the plants require cold temperatures, reduced watering, and lighting that approximates the number of hours the sun is visible. Raising the temperature or providing more hours of light than available from natural daylight can cause the bonsai to break dormancy, which often weakens or kills it.

Even for bonsai specimens that are native to the grower's location, outdoor cultivation requires specific cultivation practices to ensure successful long-term survival of the bonsai. The trees used in bonsai are constrained by the need to grow in a relatively small pot. This state greatly reduces the volume of roots and soil normally available to a freely grown tree, and brings the roots much closer to the surface of the soil than would occur in the wild. Trees in bonsai pots have much less access to water and to nutrients than they do natively, and physically confining roots changes their growth pattern and indirectly the growth pattern of the tree above the soil.

The grower has some control over the following environmental variables, and by controlling them effectively for individual specimens can ensure the health of native species grown as bonsai, and can cultivate some non-native species successfully.

- *Watering*: Different species of tree have roots with different tolerances for soil moisture. Some species tolerate continual wetness, while others are prone to rotting if the soil remains wet for long periods. A standard bonsai practice is to grow trees in a soil mixture that drains rapidly, so that roots are not allowed to be wet for long. To compensate for the relatively low water retention of the bonsai soil, water is applied frequently. The tree absorbs sufficient moisture for its needs while the water is passing through the soil, then the soil dries enough to reduce the chance of rotting. It is the grower's responsibility to ensure that watering occurs frequently enough to satisfy the bonsai with high watering requirements, while not waterlogging trees that use little water or have roots prone to rotting.

- *Soil volume*: Giving a bonsai a relatively large soil volume encourages the growth of roots, then corresponding growth of the rest of the tree. With a large amount of soil, the tree trunk extends in length and increases in diameter, existing branches increase in size and new branches appear, and the foliage expands in volume. The grower can move an outdoor bonsai from a pot to a training box or to open ground to stimulate this sort of growth. Replacing the tree in a bonsai pot will slow or halt the tree's growth, and may lead to die-back if the volume of foliage is too great for the limited root system to support. Managing the tree's available soil volume allows the grower to manage the overall size of the bonsai, and to increase vigor and growth when new branches are required for a planned styling.

- *Temperature*: Bonsai roots in pots are exposed to much greater variation in temperature than tree roots deep in the soil. For bonsai from native species, local temperatures do not generally harm the tree. But for bonsai from warmer native climates, the grower can increase the likelihood of successful cultivation either by insulating the tree from local winter conditions, or by actively increasing the bonsai temperature during the cold season. For trees from climates slightly warmer than the local one, bonsai pots can be partially buried in the ground and can be covered with an insulating layer of mulch. For trees from significantly warmer climates, warmer temperatures can be maintained in a cold frame or greenhouse, so that a relatively tender tree is not exposed to temperatures lower than it can bear. This approach may also artificially extend the bonsai's growing season, affecting watering and fertilization schedules.
- *Sunlight*: Trees generally require a good deal of sun, and most bonsai need direct sunlight during the growing season to thrive. Some shade-tolerant species of bonsai cannot thrive with too much direct sunlight, however, and it is the grower's role to site the bonsai specimens to provide the correct lighting for each type. Most bonsai will be located in an area that gets several hours of direct daylight. Shade-tolerant bonsai can be placed behind barriers (walls, buildings), sited on shaded benches or stands, or shaded by netting to reduce the impact of direct sunlight.

Repotting

An uprooted bonsai, ready for repotting

Bonsai are repotted and root-pruned at intervals dictated by the vigor and age of each tree. In the case of deciduous trees, this is done as the tree is leaving its dormant period, generally around springtime. Bonsai are often repotted while in development, and less often as they become more mature. This prevents them from becoming pot-bound and encourages the growth of new feeder roots, allowing the tree to absorb moisture more efficiently.

Specimens meant to be developed into bonsai are often placed in "growing boxes", which have a much larger volume of soil per plant than a bonsai pot does. These large boxes allow the roots to grow freely, increasing the vigor of the tree and helping the trunk and branches grow thicker. After using a grow box, the tree may be replanted in a more compact "training box" that helps to create a smaller, denser root mass which can be more easily moved into a final presentation pot.

Tools

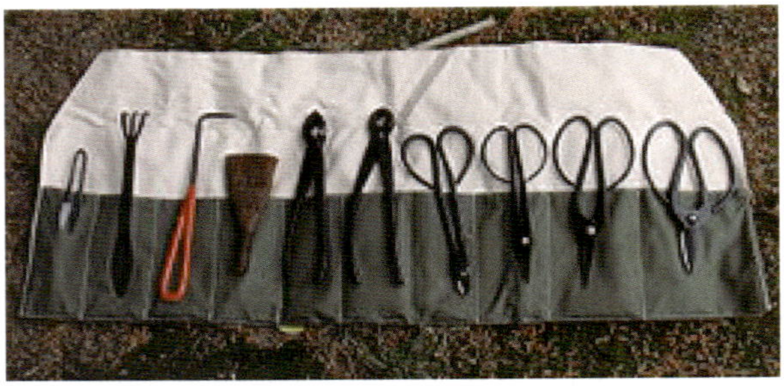

Set of bonsai tools (left to right): leaf trimmer; rake with spatula; root hook; coir brush; concave cutter; knob cutter; wire cutter; small, medium and large shears

Special tools are available for the maintenance of bonsai. The most common tool is the concave cutter (5th from left in picture), a tool designed to prune flush, without leaving a stub. Other tools include branch bending jacks, wire pliers and shears of different proportions for performing detail and rough shaping.

Soil and Fertilization

Akadama soil

Bonsai soil is usually a loose, fast-draining mix of components, often a base mixture of coarse sand or gravel, fired clay pellets, or expanded shale combined with an organic component such as peat or bark. The inorganic components provide mechanical support for bonsai roots, and—in the case of fired clay materials—also serve to retain moisture. The organic components retain moisture and may release small amounts of nutrients as they decay.

In Japan, bonsai soil mixes based on volcanic clays are common. The volcanic clay has been fired at some point in time to create porous, water-retaining pellets. Varieties such as akadama, or "red ball" soil, and kanuma, a type of yellow pumice used for azaleas and other calcifuges, are used by many bonsai growers. Similar fired clay soil components are extracted or manufactured in other countries around the world, and other soil components like diatomaceous earth can fill a similar purpose in bonsai cultivation.

Opinions about fertilizers and fertilization techniques vary widely among practitioners. Some promote the use of organic fertilizers to augment an essentially inorganic soil mix, while others will use chemical fertilizers freely. Many follow the general rule of little and often, where a dilute fertilizer solution or a small amount of dry fertilizer are applied relatively frequently during the tree's growing season. The flushing effect of regular watering moves unmetabolized fertilizer out of the soil, preventing the potentially toxic build-up of fertilizer ingredients.

Pest Management

The common pests afflicting bonsai include insects both above and beneath the soil, and infections, usually fungal. A tree grown as a bonsai is subject to the pests that affect the same species full-grown, and also to pests common to other potted plants. Most pests are species-specific, so a detailed understanding of the specific bonsai species is necessary for identifying and treating most pests. The same materials and techniques used for other affected plants can be applied to the bonsai, with some relatively minor variation. Pesticide chemicals are usually diluted more for bonsai than for a larger plant, as a regular-strength application may overwhelm the smaller bonsai's biological processes.

Location

Outdoors

Bonsai are sometimes marketed or promoted as house plants, but few of the traditional bonsai species can thrive or even survive inside a typical house. Most bonsai are located out of doors. The best guideline to identifying a suitable growing environment for a bonsai is its native hardiness. If the bonsai grower can closely replicate the full year's temperatures, relative humidity, and sunlight, the bonsai should do well. In practice, this means that trees from a hardiness zone closely matching the grower's location will generally be the easiest to grow, and others will require more work or will not be viable at all.

Indoors

Tropical and Mediterranean species typically require consistent temperatures close to room temperature, and with correct lighting and humidity many species can be kept indoors all year. Those from cooler climates may benefit from a winter dormancy period, but temperatures need not be dropped as far as for the temperate climate plants and a north-facing windowsill or open window may provide the right conditions for a few winter months.

Deadwood Bonsai Techniques

Bonsai is a Japanese art using miniature trees grown in containers, similar in some ways to the Chinese art of *penjing* and the Vietnamese art of *hòn non bộ*. The Japanese tradition of bonsai cultivation contains many specialized terms and techniques for creating bonsai and enhancing the illusion of age and the portrayal of austerity that mark a successful bonsai. Some of these methods are the deadwood techniques, which create, shape, and preserve dead wood on a living bonsai. Similar methods may exist in other traditions, but this article deals with the traditional deadwood terminology and techniques used in the Japanese practice of bonsai.

Rationale and Application

Deadwood techniques are used for reasons both practical and aesthetic. Practically, collected specimens of aged trees often have dead wood present. Dead wood can also appear on a bonsai under cultivation for many reasons, including branch die-back, pest infestation, or disease. It can be partially or completely removed by the bonsai artist, but doing so may damage the tree's overall shape or the illusion of age. If dead wood is retained, however, it must be chemically treated to preserve it and to produce the coloration of weathered wood. In addition, the dead wood usually needs to be shaped to fit the aesthetic plan for the bonsai.

Bonsai chrysanthemum

Deadwood can also be an aesthetic choice for the grower. In bonsai being developed from trees free of dead wood, it may be aesthetically useful to create some deadwood elements to enhance the illusion of age, to hide defects (such as an overlarge or misplaced branch), or to disguise the original trunk after reducing the height of a tree that is too tall. Whether freely chosen, or forced on the bonsai designer, integrating dead wood into a tree's design is a necessity for a significant number of bonsai.

Usually, deadwood techniques for branches are applied to conifers. Deciduous trees tend to shed dead branches and heal over the wound, while conifers often retain the dead limb, which naturally becomes weathered and eroded over time. Most deadwood techniques for the tree's trunk apply equally well to deciduous and to conifer bonsai, although the driftwood style (in which much of the trunk is dead) is generally restricted to conifers.

A form of *bonsai* chrysanthemum also exists, called *mum*. The cultivated flower has a life-span of about three to four years and can be kept in miniature size. The method is to use pieces of dead wood and the flower grows over the back along the wood to give the illusion from the front that the miniature tree blooms.

Deadwood Techniques

John Naka's masterpiece, *Goshin*, is on display at the United States National Arboretum. There are top jin on the leaders of several trees in this group planting.

A jin can be seen on the top left of the tree, and a shari runs up the front of the trunk from the roots on the left.

Jin

Jin is a bonsai deadwood technique used on branches or the top of the trunk (the "leader"). A jin is meant to show age, or show that the tree has had a struggle to survive. Jins are created in nature when wind, lightning, or other adversity kills the leader or a branch further down the tree. A jin requires the complete removal of bark from a given start point to the end of the branch or leader. The remaining wood dies and dries out to form the jin.

Creating a jin from the leader (a "top jin") can produce a shorter, more visibly tapered bonsai in a single step. The change in proportion can greatly improve the illusion of age in the bonsai specimen. Removing the active leader distributes vigor to lower branches, which will grow more quickly and help increase trunk diameter, reinforcing the illusion of age. A top jin also provides an aesthetic solution for a tree with two leaders, an unaesthetic shape that the designer can modify by turning one of the two into a jin.

When used on branches, the jin technique allows the grower to remove some of a bonsai's unwanted branches while increasing the illusion of age. A remaining jin may be long, with a pleasing shape bent or carved into it, or short, like the dead remains of a branch broken off near the trunk.

Uro

A uro can be seen close to the bottom of the trunk of this bonsai.

While jin appear natural on coniferous bonsai, they do not look as appropriate on most deciduous and broadleaf species. In these species, dead branches generally rot and fall off the tree. A small indentation is left where the branch used to be, and new wood grows around it forming a small hollow. Bonsai gardeners replicate this hollow as a *uro* by making a small, irregularly-shaped wound in the trunk . For example, when removing a branch from a deciduous or broadleaf species, bonsai growers often make a uro to avoid having an ugly wound healing slowly and scarring without control.

Shari and Driftwood Styles

A *shari* is deadwood on the main trunk of the bonsai. A small shari usually runs vertically on or near the front of the trunk - shari have little aesthetic value at the rear of the trunk, where they are rarely viewed and are obscured by branch growth. The shallow wound exposes a non-living portion of the trunk, which will be surrounded by living bark. The natural causes of shari include a falling branch that has ripped bark from the trunk below it, lightning damage, or trunk damage from another external source. Shari may occur naturally on a bonsai, or may be created by carving the bark.

If the amount of dead trunk (and possibly dead branches situated on the dead trunk area) is large enough, the bonsai is said to be in the *sharamiki* or driftwood style, because a large portion of the tree has the silvery, weathered look of driftwood on a beach, or ancient tree remnants on a harsh mountain landscape. "Veins" of living bark connect the roots to the live branches, but large amounts of the surrounding wood are dead, free of bark, and weathered. The dead wood may be carved into arresting shapes, to look like severely weatherbeaten tree remnants. The unusual combination of large dead areas contrasting with small signs of life is compelling regardless of the basic shape of the tree, and driftwood-style trees often do not follow the conventional bonsai styles.

Bougainvillea bonsai at the United States National Arboretum. A large sabamiki is at the bottom right of the tree

Sabamiki

Sabamiki means "hollowed trunk" or "split trunk". It gives the visual effect of a lightning strike or other severe and deep trunk damage, which has been weathered over time. Sabamiki is done by stripping bark from the trunk, then drilling or carving out the exposed wood to produce a deep wound. The hollowed area may start and end part-way up the trunk, or it may start with a wide opening at the base of the tree which tapers to closure partway up the trunk. The wound must not completely interrupt the flow of nutrients in the tree, or the branches above it will die. When the shaping work is complete, the exposed wood is treated with a preservative.

Tanuki

A Tanuki-style tree at a bonsai show.

In tanuki bonsai, a living tree is joined to an interesting piece of deadwood to create a composite in the driftwood style. The deadwood usually has the form of a weathered tree trunk, or at least its lower portion. To add living material to the deadwood, a groove or channel is first carved into it. The living tree (usually a young juniper, because of the species' vigor, flexibility, and ability to endure harsh shaping) is fixed within the channel using non-reactive nails or screws, wire wrappings, or clamps. Over time, the young tree grows into the deadwood channel, which disguises the fact that it is a separate entity. Once firmly in place, the nails, screws, or other affixing devices are removed, and the living tree is cultivated and shaped with typical bonsai techniques.

Real driftwood-style bonsai are usually not grown from the common bonsai source materials, but instead are specimens taken from the wild, and are prized but very rare. The tanuki process makes it possible to generate a driftwood-style product from much more common materials. Whether that product can be considered a traditional bonsai is open to question, as implied by the Japanese name for this technique. In Japanese folklore, tanuki, the Japanese raccoon dog, are shape-changing tricksters. Tanuki bonsai are sometimes known by the less-demeaning term "Phoenix Grafts" in the West, and many bonsai growers outside Japan consider tanuki an acceptable bonsai technique. But this technique is not currently an accepted part of the Japanese bonsai tradition, and tanuki would not be displayed at a formal Japanese bonsai show.

Tools and Techniques

Many standard shop and woodworking tools may be used in the process of creating or maintaining deadwood on bonsai. Pliers are used to grip and break off branches for jin, and are also useful for ripping off strips of bark for jin or shari. Manual tools like graving chisels, burins, and blades can carve detail into the surface of the jins or shari, since real or simulated woodgrain is an important characteristic of deadwood on a bonsai. Recent years have seen bonsai practitioners adopt powered tools for deadwood work, particularly small rotary tools for carving and grinding. When shaping is completed, a gas torch burns off remaining small shreds of wood fiber and helps raise the grain in a newly exposed piece of wood. Finally, wire brushes and sanding aids remove toolmarks and simulate weathering.

Once deadwood has been shaped to the designer's plan, the exposed area is treated with a bleaching preservative. The most common is a horticultural combination of lime and sulfur, available from many garden outlets. The preservative protects the wood from rot and pest infestation, and provides a uniform bleaching that resembles weathered, aged wood. If the lime-sulfur mixture used, it is usually colored with a small amount of dark paint to make it look more natural. Without paint pigments, the lime-sulfur solution bleaches wood to a bone-white color that takes time to weather and become natural-looking.

References

- Antimicrobial activity of phenolic constituents of magnolia grandiflora L. Alice M. Clark, Arouk S. El-Feraly, Wen-Shyong Li, Journal of Pharmaceutical Sciences, August 1981, Volume 70, Issue 8, pages 951–952, doi:10.1002/jps.2600700833
- Brooklyn Botanic Garden (2006). Mizue Sawano: The Art of the Cherry Tree. Brooklyn Botanic Garden. p. 12. ISBN 1-889538-25-6
- Brickell, Christopher (1989). The American Horticultural Society encyclopedia of garden plants. New York: Macmillan. p. 51. ISBN 0-02-557920-7
- Roh, M.S., Cheong, E.J., Choi, I-Y and Young, Y.H. (2007). "Characterization of wild Prunus yedoensis analyzed by inter-simple sequence repeat and chloroplast DNA". Scientia Horticulturae. 114 (2): 121–128. doi:10.1016/j.scienta.2007.06.005
- Nelson, Gil; Marvin, Jr Cook. The Trees of Florida: A Reference and Field Guide (Reference and Field Guides (Paperback)). Pineapple Press (FL). p. 17. ISBN 978-1-56164-055-3
- "Japanese flowering cherry". Webster's Third New International Dictionary, Unabridged. Merriam-Webster. 2002. Retrieved 2008-04-02
- Sternberg, Guy; Wilson, James; Wilson, Jim (2004). Native trees for North American landscapes: from the Atlantic to the Rockies. Portland: Timber Press. p. 268. ISBN 978-0-88192-607-1
- Blanc L, Maury-Lechon G, Pascal J-P (2000) Structure, floristic composition and natural regeneration in the forests of Cat Tien National Park, Vietnam: an analysis of the successional trends. Journal of Biogeography, 27: 141–157
- Cardoso D, Pennington RT, de Queiroz LP, Boatwright JS, Van Wyk BE, Wojciechowski MF, Lavin M (2013). "Reconstructing the deep-branching relationships of the papilionoid legumes". S Afr J Bot. 89: 58–75. doi:10.1016/j.sajb.2013.05.001
- Meirion and Susie Harries, Soldiers of the Sun: The Rise and Fall of the Imperial Japanese Army p 424 ISBN 0-394-56935-0
- Isolation and characterization of the sesquiterpene lactones costunolide, parthenolide, costunolide diepoxide, santamarine, and reynosin from Magnolia grandiflora L. Farouk S. El-Feraly and Yee-Ming Chan, Journal of Pharmaceutical Sciences, March 1978, Volume 67, Issue 3, pages 347–350, doi:10.1002/jps.2600670319
- Bush-Brown, Louise Carter; Bush-Brown, James; Irwin, Howard S. (1996). America's garden book. New York: Macmillan USA. p. 537. ISBN 0-02-860995-6
- Choi Sung-Un (2008). "Biting the cherry: Cherry blossoms and their attendant festivals herald the spring in Korea despite associations with a dark chapter with the country's history". IK-Journal. Investkorea.org. Archived from the original on 8 October 2011. Retrieved 30 November 2009
- Zion, Robert L. (1995). Trees for architecture and landscape. New York: Van Nostrand Reinhold. p. 224. ISBN 978-0-471-28524-3
- "Festa da Cerejeira em Flor 2014 celebra a Cultura do Japão em Campos do Jordão" (in Portuguese). Guia de Campos do Jordão. Archived from the original on 7 July 2015. Retrieved 7 July 2015
- Whitney, Eleanor Noss; Rudloe, Anne; Jadaszewski, Erick. Priceless Florida: Natural Ecosystems and Native Species. Pineapple Press (FL). p. 36. ISBN 978-1-56164-308-0

6

Flower Diseases and Management

Flowering plants are susceptible to a wide range of diseases, which have been discussed in detail in this chapter. Cymbidium mosaic virus, Diplocarpon rosae and Diaporthe helianthi are some of the pathogens that cause diseases in flowering plants such as orchids, roses and sunflowers. This chapter also discusses in detail the management of these diseases in flowers.

Orchid Diseases

Cymbidium Mosaic Virus

Cymbidium mosaic virus (CymMV) is a plant pathogenic virus of the family *Alphaflexiviridae*.

Cymbidium mosaic virus and the *Odontoglossum ringspot virus* (ORSV) are two of the most common viruses affecting cultivated orchids worldwide. Infected plants can have less desirable flowers or other problems, causing significant financial losses to orchid growers. The virus has not often been reported in wild orchid populations. It can be found in a wide variety of orchid genera but does not infect plants other than orchids.

Once an orchid is infected, the virus spreads throughout the infected plant in a number of weeks. Control measures may include sanitizing pruning equipment between plants. There is an ELISA test available to test for presence of the virus.

It is related to the *Narcissus mosaic virus* (NMV), the *Scallion virus X* (SVX), the *Pepino mosaic virus* (PepMV) and the Potato aucuba mosaic virus (PAMV).

Importance

Orchids are composed of nearly 200,000 species of plants with attractive flowers. They are predominantly found in wet climates. The orchid family is one of the most important plant families in respect to the ornamental flower industry. In 2005, the potted orchid industry brought in about $144 million in the United States. This makes them the second most valuable potted plant in the nation. From 1996 to 2006 there has been a 206.4% increase in potted orchid prices. Globally, Taiwan, Taipei, Thailand, the United Kingdom, Italy, Japan, Brazil, and New Zealand are among the largest importers of potted orchids. Additionally, orchids are the source of vanilla. It is the only commercially important derivation from the plants - most commonly found in the *Vanilla planifolia* species. Certain orchids are also used in homeopathic treatments.

The virus stunts the growth of orchids both through size reduction as well as lowering flower yield. CymMV has also been linked to cases of breakage in flower coloration as well as blossom necrosis. This is especially important because this causes the commercial value of the plants to be greatly reduced.

Signs and Symptoms

CymMV causes a mosaic of irregularly shaped chlorotic and/or necrotic lesions to appear on infected hosts. Additionally, infected plants will show smaller yields. Sometimes the orchids may display chlorotic rings while others will display symptoms in lesions. Sunken patches can also be observed on leaves. Lastly, certain infected plants may be symptom-less but are still viral enough to infect other neighboring plants. This is because the chlorosis and necrosis can take time to show; however, the virus can still be present on a leaf that does not display such obvious symptoms of the disease. The virus is still able to be transmitted from the plant despite its "healthy" appearance. The virus can be detected with ELISA, immunodiffusion tests, or a nucleic acid hybridization assay.

Viral Cycle

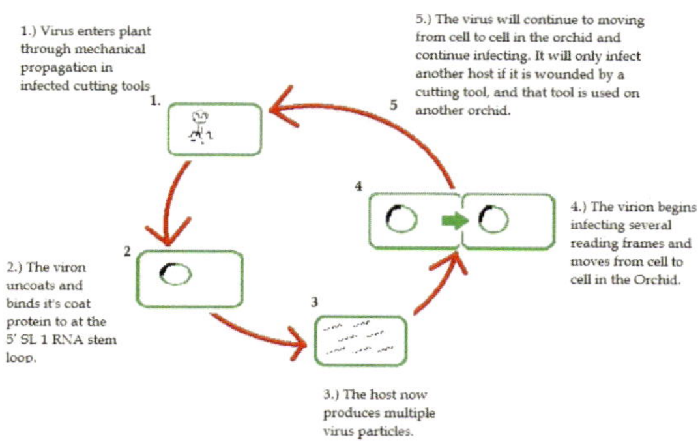

Cymbidium Mosaic Orchid Virus Life Cycle

The Cymbidium mosaic virus starts out by infecting the host through a wound on the plant. Generally this wound occurs through plant propagation using contaminated cutting tools. Once inside the orchid, the virion uncoats, binds its coat protein to the host on the 5' SL 1 RNA stem loop and begins incorporating its RNA into it. This allows for the host cell to begin producing virus particles. The virus then begins infecting reading frames such as TGB1, TGB 2, and TGB 3, each of which aid in the virus's ability to spread from cell-to-cell in the orchid. When the virus infects TGB 1, it allows for the virus to move easily through the plasmodesmata. After the virus infects TGB 2 and TGB 3, it allows for the virus to begin moving through each of the orchid cells' endoplasmic reticulum. Affecting these sites allows for the virus to move easily through each of the plant cells. Overall, the monopartite, positive, sRNA from the virion connects its 3' tail with that of the host's 5' RNA. Translations will continue to occur, and disassemble from the 5' end of the virion. With more of the virions circulating throughout the host, it only takes another wound from a cutting tool to begin infecting another orchid.

Environment

When cultivated in a greenhouse or commercially, the virus is spread mechanically if tools used in cultivation are not properly sanitized. The virus exists systemically within the plant, and persists in sap that can be spread by such methods. In the wild, the virus is most likely spread by insect vectors. A different orchid virus, the orchid fleck virus, may be transmitted by the *Brevipalpus* mite. There is also evidence that cockroaches are able to transmit CymMV.

Management

There is no way to cure a plant that has the virus. The only thing to do once a plant is infected is to destroy it. The best management solution is to prevent the spread of disease. This is accomplished through effective disinfection of tools used in cultivation, including any plastic containers and razor blades. Autoclaving, flaming, and chemical treatment with trisodium phosphate and bleach solution are traditional methods for disinfection. The use of *Streptomyces* culture filtrate, which has also been shown to disinfect mad cow disease-causing prions, is another promising disinfectant. This method is shown to be effective in removing the virus from various tools, human nails, and orchid seeds. Another possible management strategy is development of plant resistance. An attempt to transform a mutant CymMV movement protein gene into *Dendrobium* orchids was slightly successful (9 of 259 plants were resistant and expressed the marker gene); the success of transformations actually conferring resistance appears to be related to a post-transcriptional gene silencing mechanism.

Orchid Fleck Virus

Orchid fleck virus (OFV) is a non-enveloped, segmented, single-stranded (ss) RNA negative-strand virus, transmitted by the false spider mite, *Brevipalpus californicus*. OFV causes necrotic and chlorotic lesions on the leaves of many genera in the family *Orchidaceae*.

Introduction

Orchid Fleck Virus, despite its presence worldwide, only affects a small spectrum of human life. Orchids are not used for food but rather serve mainly as ornamental decoration. Therefore, only about 2 to 3 scientific reports are written about OFV each year. The whole genome of OFV has been sequenced and its six main protein products have been sequenced as well.

Much is still not known about OFV including how exactly and why vector mites travel from orchid to orchid, and more host species of flowers are being discovered annually. The more information researchers discover, the more they realize there is still much left to learn. Many orchid enthusiasts are participating in "citizen science" by posting their OFV-related findings on international horticulture blogs and forums.

Virion Morphology

OFV was first described as bacilliform but depending on the method of preparation, OFV can appear bullet-shaped or bacilliform. On average, OFV is 40 nm in diameter and between 100 and 150 nm long. Each viral particles is organized into a tight 25 turn helix, with a pitch of about 4.5 nm.

Genomics

OFV contains two ssRNA molecules, RNA1 and RNA2, of 6413 and 6001 base pairs, respectively. GenBank contains the whole sequenced genome of OFV. RNA1 (GenBank AB244417) codes for five proteins whereas RNA2 (GenBank AB244418) only codes for one. Both strands possess open reading frames (ORF), which are read in the negative sense.

RNA1

- ORF1: Nucleocapsid (N) protein
- ORF2: Phosphoprotein (P)
- ORF3: Proteins involved in viral cell-to-cell movement
- ORF4: Matrix (M) protein
- ORF5: Glycoprotein (G)

RNA2

- ORF6: Polymerase (L) protein (RNA-dependent RNA polymerase)

Pathophysiology

Chlorotic and necrotic flecks, spots, and/or ringspots, as well as yellow flecks or spots are all symptoms of an OFV infection. Studies have also shown that OFV may prevent the propagation of other viruses in an already OFV-infected plant.

Vector

The false spider mite, *B. californicus* serves as the major vector for OFV. *Brevipalpus* mites go through four distinct, active life stages, each separated by nonmotile chrysalis stages. The protonymph, deutonymph, and adult stages can infect their host plants with OFV, whereas the larval stage is not infectious. Even after three weeks of incubation of an OFV-positive mite on an OFV-resistant plant, *B. californicus* proved to still be infectious, showing that OFV is persistent.

Hosts

OFV is able to naturally infect around 50 different species in 31 genera, all belonging to the *Orchidaceae* family. 25 other species from 11 non-orchid families have been infected through sap transmission or artificial viral inoculation.

Life Cycle

Studies have not shown whether or not OFV actually replicates within *B. californicus* but electron microscopy has revealed an intricate viral life cycle within the host cells.

Viral ssRNA is replicated and transcribed into mRNA in the host cell's nucleus. Viral mRNA is then exported out of the nucleus into the cytoplasm where it is translated into viral protein by the host's ribosomes. The viral proteins then reenter the nucleus where they aggregate into a viroplasm. There, the various viral structural proteins assemble with both strands of ssRNA to form complete OFV particles. These particles often cluster in between the inner and outer nuclear membranes, causing visible projections which often evaginate into cytoplasmic vesicles. Electron microscopy has revealed clusters of viral particles positioned perpendicular to the inner nuclear membrane, the endoplasmic reticulum, as well as the aforementioned cytoplasmic vesicles, forming distinctive "spoked wheel" structures.

Effects on Fitness

Infected orchids don't bloom as well as healthy ones, affecting efficacy of pollination and fertilization. Also, the orchids that do bloom look lifeless making them less attractive on the cut flower market.

Epidemiology

Cases of Orchid Fleck Virus or OF-like Viruses have been reported in Australia, Brazil, China, Columbia, Costa Rica, Denmark, Germany, Japan, Korea, South Africa, and the United States, i.e. every continent except for Antarctica.

Due to the fact that viruses depend on their host cell for replication, OFV cannot be cultured independently. However, two non-orchid indicator hosts (plants used in research that show characteristic symptoms of specific viral infections) *C. quinoa* and *T. expansa* are commonly used for viral inoculation and isolation.

There are no known pathogens of OFV itself but its vector, *B. californicus* has a symbiotic relationship with bacteria of the genus *Cardinium*. The symbiont is the cause of the mites' thelytokous method of reproduction (where females are produced from unfertilized eggs) and the explanation for the absence of male *B. californicus* mites.

Diagnosis

Thin tissue samples from plants with visible symptoms of OFV can undergo:

- Electron microscopy to visualize virions and complete viral particles
- Serological analysis to identify and isolate specific viral proteins
- Reverse Transcription-Polymerase Chain Reaction to identify viral RNA

Prevention

Methods for preventing the spread of OFV among separate plants:

- Ensure seedlings are virus-free
- Beef-up quarantine measures
- Eliminate sources of infection (mites or other infected plants)
- Ensure proper environment for cultivation
- Work towards developing OFV-resistant plants through Genetic Engineering

Treatment

Once a plant is infected with OFV, it is unclear whether pruning visibly infected tissue will cure the plant of the virus. It is also unclear whether infected plants produce seeds containing viral particles.

Human Relevance

Orchids and other tropical flowers are extremely important to the agricultural economy of many Southeast Asian countries. The Vanda 'Miss Joaquim' orchid hybrid is Singapore's national flower. According to the Orchid Society (citation needed), the world market for orchids is worth more than $1 billion with Japan and the United States leading the way in 2010. In the US in 2005 produced around $144 million worth of orchids.

Rose Diseases

Diplocarpon Rosae

Diplocarpon rosae is a fungus that creates the rose black spot disease. Because it was observed by people of various countries around the same time (around 1830), the nomenclature for the fungus varied with about 25 different names. The asexual stage is now known to be *Marssonina rosae*, while the sexual and most common stage is known as *Diplocarpon rosae*.

Diplocarpon rosae over seasons as mycelia, ascospores, and conidia in infected leaves and canes. In the spring during moist, humid conditions, ascospores and conidia are wind-borne and rain-splashed to newly emerging leaf tissue. Upon infection, disease progresses from the lowest leaves upward, causing defoliation and black spots on leaves.

Diagnosis

The black spots are circular with a perforated edge, and reach a diameter of 14 mm. Badly affected plants, however, will not show the circular patterning, as they combine to cause a large, black mass. The common treatment of the disease is to remove the affected leaves and spray with antifungal solutions. Some stems of the roses may become affected if untreated, and will cause progressive weakening of the rose.

Disease Cycle

Diplocarpon rosae tends to overwinter in both lesions of infected canes and fallen foliage. Conidia are produced in the diseased stem tissues and dispersed via water—most commonly by rain or wind—into the openings of leaves in the spring season. The conidia then produce germ tubes (and sometimes appressoria) to penetrate the tissues of the leaves. Mycelia develop on the underside of the leaf cuticle and lesions appear. As these lesions appear, acervuli continuously produce conidia asexually as long as the climate remains optimally wet and warm. These conidia can then be dispersed to new uninfected leaves as a source of secondary inoculum, adding more cycles of infection. Once defoliation occurs in the fall season, the hyphae of the *Diplocarpon rosae* invade the dead leaf tissue and form pycnidia lined with conidiophores under the old acervuli. The pycnidia then overwinter in the lesions of infected tissue and burst in the spring, releasing conidia to be dispersed by water and effectively completing the disease cycle. *Diplocarpon rosae* also has a sexual stage, although this is rarely observed in North America due to unfavorable environmental conditions. In this stage, the sexual spores (ascospores) are formed in the apothecium. If the weather conditions are favorable for the formation of ascocarps, the apothecia that contain asci can be observed in the spring. However, this rarely occurs, and the fruiting bodies are typically filled with conidia that enable the asexual life cycle of the pathogen to occur.

Environment

Diplocarpon rosae typically favor environments with a warm and wet climate. Conidiospores involved with infection are only dispersed via water, making the disease most active in the late spring and early fall seasons, or other periods that experience similar climate conditions. The development of the Black Spot disease itself is ideal at temperatures ranging from 68-80° F. It is important to note that no infection will develop if the leaf surfaces dry out within 7 hours of the initial conidial germination. Similarly, temperatures above 85° F also halt the spread of disease.

Treatment

Removing infected leaves from the plant and fallen leaves from the ground will slow the spread of the infection, as does avoiding wetting the leaves of plants during watering. An infected plant can be removed from the area, which will slow the spread of infection to other plants, but this often is not desirable. Fungicides, such as mancozeb, chlorothalonil, flutriafol, penconazole, or a copper-based product, applied upon new leaf emergence or first appearance of black spot, can be used to control the disease. If a more natural and nontoxic approach is desired, diluted neem oil is effective both against black spot and as an insecticide against aphids. It is usually necessary to repeat the spraying at seven- to 10-day intervals throughout the warmest part of the growing season, as the fungus is most active at temperatures from 24 to 32 °C (75 to 90 °F).

Importance

Black Spot of rose is the single most impactful disease of roses globally. Every year around 8 billion flowering stems, 80 million potted plants and 220 million garden rose plants are sold commercially. All species of roses (*Hulthemia, Hesperrhodos, Platyrhodon* and Rosa) are affected by black spot disease. The disease is found everywhere roses are planted, typically in epidemic proportions. The water-borne dispersal methods allow it to infect a plethora of plants every growing season and increase the overall incidence of disease. Although *Diplocarpon rosae* does not kill the rose itself, it is known to completely defoliate the leaves of the rose plant. This is a huge issue when dealing with such an aesthetically commercialized crop such as the rose. Additionally, the weakened rose plant will become more susceptible to other pathogens and disease following infection.

Podosphaera Pannosa

Podosphaera pannosa is a plant pathogen. It produces a powdery mildew on members of the rose family.

Podosphaera pannosa on *Rosa* sp. leaf

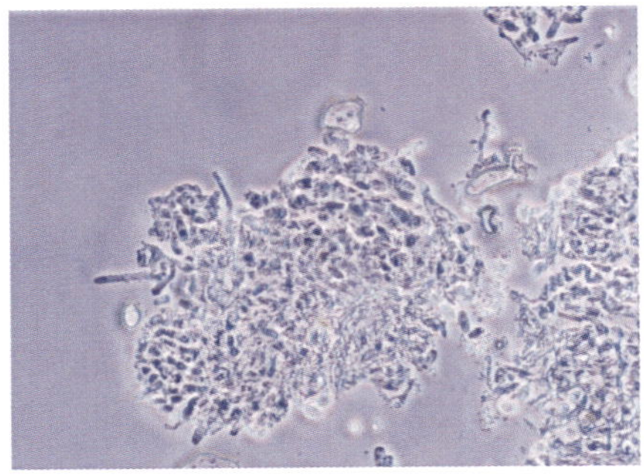

Podosphaera pannosa

Summary

Rose powdery mildew [also known as 'Weeping Mildred'] is caused by the fungus *Podosphaera pannosa*, a member of the Ascomycete fungi. It infects a wide variety of roses, but especially those grown in dryer climates as the fungus has the rare characteristic that not only does it not need water to germinate and reproduce, it can be inhibited by it.

Disease Cycle

The disease cycle of rose powdery mildew starts when the sexual spores, ascospores, of the pathogen survive the winter in a structure composed of hyphae called an ascocarp. The specific ascocarp is a chasmothecium, or cleistothecium, and has a circular shape to it. Under the right conditions the chasmothecium will break open to reveal the asci, which are long tube-like structures containing the ascospores. These ascospores are then released and spread by wind, insects, and rain until they land on a susceptible rose for a host and land, attach, and germinate on the plant. They will also produce condia, the asexual spores of *Podosphaera pannosa*, which will spread throughout the summer. It is these long chains of white conidia which give the fungus its characteristic "powdery" appearance. Late in the year as the plant is dying cleistothecia will again form when the ascogonium receives the nucleus from the antheridium.

Environment

Optimal conditions for rose powdery mildew are 16-27 degrees Celsius, with the optimal temperature for fungal growth at 23 degrees Celsius in a shaded area. They also do not need water to germinate and infect the rose. In fact, if there is too much water present on plant surfaces for a prolonged period of time the fungal growth is inhibited and the spores can actually die. Rose powdery mildew can also grow in any conditions where roses can grow and has been found everywhere from China to the United States.

Hosts, Signs, and Symptoms

A wide variety of rose species are susceptible to powdery mildew. In light of this it is more practical to discuss the rose varieties that are resistant as opposed to those that are susceptible. Two varieties that have been found to show resistance to rose powdery mildew are "Paul's Pink" rose variety and the "Magic" rose variety. Other research has shown that many chestnut rose (*Rosa roxburghii*) varieties are also resistant to powdery mildew. *Rosa sterilis, Rosa kweichowensis, Rosa laevigata, Rosa lucidissima,* and *Rosa chinensis* have all been shown to be resistant to powdery mildew. *R. multiflora var. multiflora* and *R. multiflora var. cathayensis* have all been shown to be susceptible to rose powdery mildew. Symptoms caused by the rose powdery mildew can be a dwarfing of the growth of the plant, or the twisting and deforming of leaves, but more noticeable is a sign of the disease, which is the white condia, the "powder" that appears on the plant surfaces, such as leaves, shoots, flowers, and buds. The fungus may grow on both new and old leaves, but is generally more concentrated on the underside of the leaf.

Management

Effective management of rose powdery mildew begins by using resistant varieties of rose, but it can also be managed through the use of fungicides, or by planting in sun since rose powdery mildew prefers the shade. In fact, increasing the exposure of rose powdery mildew from 18 to 24 hours of light per day reduced the production of conidia, the asexual spores of the fungus, by as much as 62%. There are a variety of fungicides that have proven to be effective. Examples are myclobutanil, azoxystrobin, triadimefon, and thiophanate-methyl formulations Chemical fungicides are not always necessary, however, it is possible to use more environmentally-friendly solutions such as a water-vinegar spray, or mixtures of baking soda and insecticidal soaps Recent studies have also shown that using a planting medium which includes silicon can also reduce the occurrence of powdery mildew by as much as 57%.

Importance

Powdery mildew affects more the 7600 species of hosts worldwide, including subsistence crops Although rose powdery mildew will most directly affect the rose connoisseur, it is part of this larger family of powdery mildews, which can affect the crops used for food and survival in many countries, thereby having economic and human impacts beyond that of an unsightly rose bush. Research shows that total yield loss from powdery mildew on cereal crops alone can vary anywhere between 2 and 30% depending on the host and the environmental conditions. Additionally, the wholesale value of roses annually exceeds $100 million in the United States, so the national economic impact from the flower industry cannot be ignored As explained by an old gardener's adage 'Weeping Mildred! She's no friend of the gardener and certainly no friend of me'.

Sunflower Diseases

Diaporthe Helianthi

Diaporthe helianthi is a fungal pathogen that causes *Phomopsis* stem canker of sunflowers. In sunflowers, *Phomopsis helianthi* (teleomorph =*Diaporthe helianthi*) is the causative agent behind stem canker. Its primary symptom is the production of large canker lesions on the stems of sunflower plants. These lesions can eventually lead to lodging and plant death. This disease has been shown to be particularly devastating in southern and eastern regions of Europe, although it can also be found in the United States and Australia. While cultural control practices are the primary method of controlling for Stem Canker, there have been a few resistant cultivars developed in regions of Europe where the disease is most severe.

Leaf lesion proceeding along major leaf vein

Host and Symptoms

Stem canker on sunflower is caused by the fungal pathogen *Phomopsis helianthi* (teleomorph = *Diaporthe helianthi*). There are many "Phomopsis" species that have a broad host range, but "Phomopsis helianthi" is the primary pathogen that infects sunflower, although it has also been seen to infect other, secondary hosts. Some "Phomopsis" species have been found to be able to cause disease on several host genera. Sunflower hosts that found in high densities are particularly susceptible to attack by the pathogen due to the microclimate formed in crowded areas. This microclimate has a high humidity and allows for infection to proceed rapidly. There has also been recent molecular evidence to suggest there are several genetically distinct species of *Phomopsis* that can cause disease on the sunflower hosts.

Initial symptoms of *P. helianthi* infection can be found at the attachment of petiole to the stem on the lower portion of the plant. Initial infection will cause brown necrotic regions with a chlorotic border. Eventually these regions will grow into larger, brown to tan lesions and eventually into large cankers. Cankers are found on the stem and will have a sunken border. Cankers can be long, up to 6 inches, but do not appear until flowering. Other early symptoms can include necrosis and bronzing of intervenial areas of leaves at or above the stem lesion. Wilting and death of leaves above the canker also occurs. Leaves above the site of inoculation can also show symptoms of being water soaked. This is sometimes accompanied by leave distortion. *Phomopsis* stem canker is also characterized by extensive pith degradation to the point where the stalk is easily crushed under moderate thumb pressure. Wilting is also a typical symptom, and extensive wilting can eventually lead to lodging later in the season. *Phomopsis* also produces mycotoxins, which can cause additional symptoms such as premature ripening or early plant death.

Disease identification is facilitated by observation of fruiting structures. Perithecia are found on cortical tissues and will produce globular to sub-cylindrical asci with two-celled ascospores. Interestingly, it has been observed that isolates from stem canker in the United States will readily form perithecia when grown in culture, but those from Europe will not. This suggests that there may be several species or biotypes of *Phomopsis* that cause stem canker worldwide. Pycnidia, the asexual fruiting bodies, can be detected on the stem and leaf lesions throughout the disease cycle. They are embedded in the tissue, dark brown and globular. The conidia come in two types, but the most common is beta conidia that are threadlike, hyaline and can be curled or straight.

Petiole infection leading to stem infection

Disease Cycle

Stem canker on sunflower has both a teleomorphic and anamorphic stage. The anamorphic stage is given the name *Phomopsis helianthi.* When the teleomorphic stage was discovered, it was given the name *Diaporthe helianthi.* Infection is initiated onto the leaves by windblown or rain splashed ascospores that were released from overwintering perithecia, the telomorphic fruiting body. The infection begins at lower leaves and spreads through the intercellular spaces and terminal veins throughout the vascular system. The leaves that contain the original inoculum will quickly wilt and die but by then the infection has already entered the stem where it persists. Pycnidia, the anamorphic fruiting body, will form in the infected tissue and release two types of conidia – alpha and beta. Alpha conidia are oval to fusoid in shape while beta conidia are threadlike. Both conidia are produced, but the proportion of alpha to beta may change depending on environmental conditions. Beta conidia are most frequently produced. At the end of the season, perithecia formed on leaves and other diseased tissues will remain in the debris and survive for the next season. The perithecia can survive in debris for up to five years and *Phomopsis* has also been shown to be dispersed through seed.

Lodging due to stem infections.

Environment

This disease is most severe in southern and eastern Europe but several cases have been reported from Argentina as well as USA. The overwintering structures (perithecia) allows this fungus to survive in unfavorable conditions. Ascopsores from these perithecia are released during the growing season and are spread by wind and rain splash throughout the season. Relative humidity of 90% and temperatures ranging from 20-24°C favors the disease development. Abundant rainfall during the budding to flowering stage promotes the infection more than the temperature.

Management

The fungus produces overwintering structures which makes it difficult to manage. Proper treatment (burning or burying deep under the ground) of the plant waste in fall and crop rotation can decrease disease incidence and severity. Some *Phomopsis* resistance hybrids are available in Europe where the disease is particularly severe. There are no commercial hybrids available in US which are resistant. There are several methods that can be employed to control this disease, some of which are discussed below:

Genetic Control

In early 1980s all the commercial hybrids available were completely susceptible. Only a couple of varieties like 'Yubileynaya 60' and 'Progress' were somewhat tolerant with only 5% of plants without disease. It was found out that resistance to this pathogen is polygenic rather than controlled by single dominant gene. After several years of study and research, there were some hybrid varieties developed in Europe that were indeed resistant to *Phomopsis*. Studies have shown that the genetic mechanisms of resistance correspond to a few genes with additive effects.

Chemical Control

Fungicides were also used to control this disease. Although fungicides must be applied before symptoms appear, typically during flowering stage. Treatments after appearance of symptoms were proven to be ineffective. Application of fungicide reduced the severity of the disease and increased the yield but was unable to reduce the number of infected plants significantly. To this date there are no registered fungicides in U.S. for control of this disease.

Cultural Control

Cultural control serves a key role in the control of this disease. Controlling crop densities, Nitrogen fertilization and crop rotation may help decrease the incidence of this disease. In 2010, Debaeke and Moinard showed that the early *Phomopsis* attacks were related to changes in microclimate resulting from crop management practice or cultivar architecture.The later attacks were more dependent on surface area of leaf available and air movement in the canopy provided the inoculum is present and microclimate is favorable to germination of spores. It was found that the pathogen was not found if the infected stems were incorporated >5cm below ground.

There are currently no biological control strategies developed for this disease.

Importance

Phomopsis stem canker of sunflower is a serious disease caused by *Phomopsis helianthi* (sexual stage = *Diaporthe helianthi*), it was first observed in Europe in the late 1970s followed by U.S. in 1980s. Some of the earliest reports of this disease came from certain areas of Ohio State. After careful studies and examinations, they were able to detect pycnidia of *Phomopsis* spp. Emergence of this new disease became more and more apparent with increased number of reports from all over the world including Rumania in 1985, Hungary in 1983, France in 1985, and Iran in 1988. Relatively recently, Stem cankers were observed on sunflower plants growing in a field in Illinois. This is also becoming an emerging problem in Australian sunflower plants with yield losses reaching up to 40%. Since the fungus is highly specific to sunflower, it will be difficult to find it in areas without major sunflower production. Disease is most severe in high temperatures and high rainfall leading to significantly high yield losses. NSA crop surveys show significant yield losses in Minnesota (22% infected plants in 2007)as well as North Dakota with an incidence of over 40% plants infected. Yield losses usually go higher with lodging.

Plasmopara Halstedii

Plasmopara halstedii is a plant pathogen infecting sunflowers. The species is one of many pathogens commonly referred to as downy mildew. *P. halstedii* originated in North America.

P. halstedii oospores produce a thin wall which are resistant structures, sexually produced that are essential for its continuation. After entering an area, the eradication of the pathogen is difficult due to the formation of oospores, which can remain viable in soil for many years.

Hosts and Symptoms

Plasmopara halstedii is an obligate biotroph that attacks the flowering plants of the Asteraceae family, found to infect the genus' *Helianthus, Bidens, Artemisia, and Xanthium*. The pathogen has the strongest impact on *Helianthus,* degrading flower yields in the species *H. argophyllus, H. debilis, H. petiolaris and H. annuus*. H. annus, the common sunflower, is also the most common host of *P. halstedii. Xanthium strumarium*, the common cocklebur, and *Ambrosia artemisiifolia*, or ragweed, have been shown to act as significant wild hosts.

P. halstedii causes significant yield losses due to the production of infertile sunflowers. Infertility due to P. halstedii is a result of sporulation on the flowering bodies as well as seed damping off due to root infection. Other symptoms include plant stunting, chlorosis, root browning and alteration of secondary metabolism of infected plants.

As an oomycete, *P. halstedii* releases motile zoospores as secondary inoculum. In these cases, secondary infection symptoms are much less severe than primary symptoms. A sign of the disease is the "typical downy appearance" resulting from sporulation the plant surface.

Detection

Systemically infected sunflower plants may have some degree of stunting and the leaves show pale green or chlorotic mottling which spreads along the main veins and over the lamella.

Disease Cycle

P. halstedii is a plant pathogenic oomycete, capable of overwintering in soil due to survival structures called oospores. For this reason, *P. halstedii* is a soil borne pathogen infecting the roots of the host plant. Oospores have the potential to live in soil up to 8 years, while oospore germination takes 10–30 days. Germination length depends on environmental condition and typically occurs in the spring. The germinating oospores form sporangia that release motile zoospores as a secondary inoculum.

After primary infection, zoospores serve as a main source of inoculum throughout the rest of the season. Zoospores germinate in about two hours and have two means of infection: direct penetration of the root through the use of an appresorium or infection due to an already present wound. Hyphae extend throughout the intercellular space, forming parasitic haustoria. The pathogen breaks through the surface of leaves and flowering structures through stomata. Asexual reproduction structures, called zoosporangiaphores, form at these sites. These structures release zoosporangia containing zoospores to other plants, the primary means of dissemination.

Sexual reproduction be either homothallic or heterothallic. Homothallic reproduction is characterized by the fusion of the asexual oogonium and antheridium. This fusion leads to the formation and release of sexual oospores, the primary inoculum for the next season. Heterothallic sexual reproduction is the fusion of sexual cells from two separate organisms, leading to "outcrossing".

Management and Control

Once the pathogen has been detected in an area, management is essential, as *P. halstedii* is nearly impossible to eradicate. Between long surviving resting spores and high levels of secondary inoculum, *P. halstedii* can infect anywhere from 50% to 95% of sunflower yields in a single season.

Resistant sunflower strains are available, as two types of dominant major resistance genes have been identified, denoted as *Pl*. Type 1 resistance lacks infection above the base of the hypocotyl. Type 2 is characterized by a weak infection, with sporulation never reaching the upper region of the host. Resistant genes have been overcome in the past as the pathogen evolves and mutates, suggesting that resistance may not give the best results. However, due to the constant search for resistant sources from wild Helianthus, the gene pool of cultivars is frequently being used to produce new, commercially resistant strains.

Seed treatment has been shown to be effective in controlling the disease, as the establishment of *P. halstedii* in an area of soil is nearly irreversible. The compounds metalaxyl and oxadixyl have been shown to protect seeds in the case of infection, and treatments containing these compounds are commercially available. Fungicides seed dressings of this nature have also been shown to give significant control over the pathogen. However, some strains of *P. halstedii* have begun to show resistance to metalaxyl-based fungicides, reported in multiple countries. Alternate forms of resistance are being developed for the future, including the use of biological antagonists and defense-related proteins.

Australia is the leader in pest control of *P. halstedii*. In Australia, any imported seed is subject to hot-water treatment, fungicide dusting and monitoring for up to 3 years.

Common Disease Pathogens

Phoma Clematidina

Phoma clematidina is a fungal plant pathogen and the most common cause of the disease clematis wilt affecting large-flowered varieties of *Clematis*. Symptoms of infection include leaf spotting, wilting of leaves, stems or the whole plant and internal blackening of the stem, often at soil level. Infected plants growing in containers may also develop root rot.

Taxonomy

The asexual stage (anamorph) of the fungus was first described by the German botanist and mycologist Felix von Thümen in 1880 as *Ascochyta clematidina*. Based on new scientific insights into the differences in spore formation between species, it was reclassified as *Phoma clematidina* by the Dutch mycologist Gerhard Boerema in 1978.

Genetic sequencing has suggested that *Phoma clematidina* is heterothallic which means that two compatible strains (mating types) of the fungus would need to come together under the right environmental conditions to produce a sexual stage (teleomorph). Both mating types of *Phoma clematidina* are known to occur in Europe, and yet no sexual stage (which is most likely to be a *Didymella* species) has ever been described.

Molecular phylogenetic analyses have revealed that some fungal isolates recovered from wild *Clematis* species, previously identified as *Phoma clematidina*, are in fact two closely related species of *Phoma* and *Ascochyta* with the sexual stages *Didymella vitalbina* and *Didymella clematidis* respectively. The latter fungus has been successfully used as a biological control agent of *Clematis vitalba*, which is seen as an invasive plant in New Zealand. Unlike *Phoma clematidina*, the two closely related *Didymella* species (and their anamorphs) have not been found on large-flowered *Clematis* varieties.

The previous misidentification of these species means that some literature, particularly that on the biological control of *C. vitalba*, referring to *Phoma clematidina* is actually describing work on *Didymella clematidis* and its *Ascochyta* anamorph.

Biology

The biology of *Phoma clematidina* is now reasonably well understood thanks to research into clematis wilt at the University of Canterbury in New Zealand and the University of Derby and ADAS in the UK.

A full overview of its life cycle was first produced in 1999. Within a nursery or garden, the fungus is mainly spread through splash dispersal of its asexual spores (conidia) which are formed in bulbous fruiting bodies called pycnidia. Pycnidia can be formed on any infected part of the plant showing symptoms, including leaves, stems and roots. The pycnidia exude the spores in a sticky mass which is splash dispersed to other plants nearby during rain or irrigation. Contact, such as with pruning tools, may also spread the spores and, in addition, it is suspected that certain insect species may act as contact vectors. Movement of infected plant material forms an important method of spread to previously unaffected nurseries or gardens.

Unlike many other plant pathogens, *Phoma clematidina* is not a biotroph which means that it can infect plants but does not need them to survive. The fungus can live saprophytically on dead plant material or organic matter as well. In addition, it forms thick-walled resting spores (chlamydospores) which increase its survival in plant debris and soil during unfavourable conditions.

Disease Development

Many publications report *Phoma clematidina* to be a wound pathogen. However, although wounding may aid infection, scientific trials have shown that, in susceptible *Clematis* varieties, the fungus can cause extensive leaf spotting and wilt without any previous damage to plant tissues.

When spores of *Phoma clematidina* land on a leaf of a susceptible plant under the right environmental conditions (moisture, temperature 15 to 25 °C), they will germinate within 6 hours and infect leaf tissues directly or via leaf hairs (but never via stomata). Initial symptoms are small brown lesions which then rapidly spread, eventually killing the whole leaf. Unless the plant abcises the leaf in reaction to the infection, the fungus will spread via the leaf stalk to the node of the stem.

Infection may also occur directly into the stem. This is particularly common at or below soil level where the right amounts of moisture are more likely to persist. Once in the stem, *Phoma clematidina* destroys the internal tissues causing a distinctive, localised black discolouration. Such stem infections can be hugely destructive as they block the plant's vascular system and will lead to the wilting and death of all plant mass above the site of entry, no matter how substantial.

Relatively recently, it was discovered that *Phoma clematidina* can also infect the roots of *Clematis* and be a cause of black root rot. In practice, this has been found to be a particular issue in containerised plants.

Host Susceptibility

Many gardening publications express a view on the susceptibility of different *Clematis* varieties and species to clematis wilt based on observations in practice, but very few comment on their susceptibility to disease caused by *Phoma clematidina* in particular compared with other causes of wilt.

Scientific trials have shown that large-flowered varieties are especially susceptible to disease caused by *Phoma clematidina* in line with their vulnerability to wilt observed in practice. It is believed that hybrids which have *Clematis lanuginosa* in their ancestry are most susceptible to wilt. Such hybrids include many cultivars in the early flowering, large-flowered group (such as *C.* 'Nelly Moser' and *C.* 'Elsa Späth') and, to a lesser extent, cultivars in the late flowering, large-flowered group (such as *C.* 'Jackmanii' and *C.* 'Perle d'Azur').

Clematis viticella and related varieties are also susceptible to infection with *Phoma clematidina*, yet are reported to wilt less often than large-flowered varieties, possibly because they are very vigorous and outgrow infections quite rapidly. Most other cultivated *Clematis* species and their varieties are largely resistant to both infection with *Phoma clematidina* and wilt in general.

Phoma clematidina has sporadically been reported from species of *Clematis* growing in the wild, including *Clematis orientalis* in Russia and *Clematis pubescens* in Australia. Reported isolations from wild *C. vitalba* in Europe have proven to be *Didymella vitalbina* rather than *Phoma clematidina*.

Lilac Witches'-broom

Lilac witches'-broom (LWB) is a disease of lilacs caused by the Phytoplasma *Candidatus Phytoplasma fraxini*. This phytoplasma was first identified as the causal agent of Ash Yellows and has since been attributed to both diseases.

Symptoms of the LWB phytoplasma include witches' brooms, shortened internodes on new growth, twig dieback, overall loss of vitality and premature death. Detection is primarily performed based on physiological symptoms in the host plant, although a new identification based on direct gene sequencing has been confirmed. There is no known cure for this disease so management strategies involving culling and proper disposal of infected specimens is encouraged.

Witches brooms at base of declining lilac

The disease is believed to be spread by means of an insect vector, most likely of the order hemiptera. Certain cultivars of lilacs appear to be more susceptible, notably those of the Preston Lilac Series developed at the Central Experimental Farm in Ottawa Canada. Recent research performed there suggests heat treatment of scion material as an effective means of sterilization.

Ovulinia Azalea

Ovulinia azaleae is a plant pathogen affecting azaleas and rhododendrons.

Ovulinia Petal Blight

First reported in the 1930s on the Belgian-Indica hybrids, petal blight can attack most azaleas under favorable weather conditions. Infected flowers first exhibit small spots, about 1 mm in diameter, which appear water-soaked. The spots enlarge rapidly and become very slimy, causing entire petals to become slimy and limp, usually within 2–3 days after initial infection. Infected areas of flowers soon become tan or light brown, and eventually entire flowers turn prematurely brown. Infected flowers dry and generally cling to the plant longer than uninfected flowers do. One to 20 (commonly two to five) small, black sclerotia (a fungal resting stage) form on an infected flower 6–8 weeks later. They appear first as small white areas on the dried petals and slowly become dark as they mature.

Caused by the *Ovulinia azaleae* fungus, the disease develops during periods of moist weather at flowering time. Early- and late-flowering cultivars or species tend to escape the disease. The fungus overwinters as sclerotia on diseased petals adhering to plants, on the soil surface, or in leaf mulch under the bushes. At the beginning of the host's blooming period, the sclerotia germinate and in 3–5 days give rise to small stalked, cup-shaped fruiting bodies filled with spore sacs, which are discharged with sufficient force that they strike the lower blooms above the sclerotia or are carried in air currents to nearby blooms on adjacent plants.

The rate of infection is higher when periods of frequent precipitation and warm weather coincide with flowering. Heavy dew or extended periods of misty weather are particularly favorable. Low temperatures at the beginning of the flowering period and dry conditions toward the end of it most likely reduce the infection potential.

Spraying the infested soil with fungicides and removal of the leaf litter beneath infected plants have not been reliable methods of disease control. Sprays applied at or just before bud opening have offered the best means of control. In general, sprays of triadimefon or benomyl applied at weekly intervals throughout the flowering period have controlled the blight and inhibited sclerotium formation. Making sure diseased or disease-carrying plants are not introduced into the growing area reduces the disease potential.

Pestalotia Rhododendri

Pestalotia rhododendri is a plant pathogen infecting azaleas and rhododendrons.

Important diseases: Pestalotiopsis tip blight of conifers, Gray leaf spot

Pestalotia is primarily a secondary pathogen. It is saprophytic on dead and dying tissues and is weakly parasitic infecting wounds under moist conditions.

Tips of conifer branches particularly Leyland cypress, arborvitae and juniper turn brown to grayish in color. Infected bark may be covered in fungal fruiting structures giving the tissue a black sooty appearance. Leaf spots tend to be tan to gray and are often the result of previous damage such as freeze injury, scorching or mechanical wounds.

Dark, disc or cushion-shaped acervuli are formed under the plant epidermis which then splits open revealing the fruiting structures. Conidia are produced on short simple conidiophores within the acervulus.

Conidia are multi-celled with usually three darkly pigmented center cells and clear pointed end cells. Conidia are ellipsoid or fusoid (football-shaped). A diagnostic feature is the two or more clear, whisker-like appendages arising from the end cell

Phytophthora Cactorum

Phytophthora cactorum is a plant pathogen that causes root rot on rhododendron and many other species.

Hosts, Symptoms, and Diagnosis

Phytophthora cactorum has an extremely wide host range, and can infect over 200 species or 160 genera of trees, ornamentals, and fruit crops. In general, *P. cactorum* is capable of infecting both young and old plants, and causes root rots and crown rots of the many genera it infects. Although the symptoms this pathogen produces varies between the types of organisms it infects, in general disease occurs during periods that are both wet and warm. Furthermore, most infections are caused by zoospores entering the plant through wounds.

On older trees, *P. cactorum* causes the formation of sap exuding dark colored cankers on the trunks of trees, as well as leaf size and number reduction, chlorosis, and dieback of branches. The diagnosis of a *P. cactorum* infection of trees, is based on the identification of symptoms, in particular the oozing cankers, and confirmation of symptoms in a diagnosis lab or utilization of a field ELISA detection kit. *Phytophthora cactorum* can be a major problem in apple orchards, as it can cause crown, collar, and root rots in apple trees. When infecting apple trees, the organism can attack through wounds either above or below the soil line, impairing phloem and root function, and causing stunting, foliar disorders, and death after several years. Furthermore, because the pathogen causes damage to the phloem of the tree, one diagnostic method is to check for necrotic phloem tissue at the base of the tree which will be orange to red-brown in the early stages and dark brown in the later stages of infection.

A good example of *Phytophthora cactorum* causing foliar disease, is *Phytophthora cactorum* on ginseng. Foliar disease of ginseng usually occurs during May and early June, causing the leaves to become transparent and papery. Ginseng foliar infection occurs through the rain splash dispersal of spores from the soil onto above ground wounds. Once infected, *Phytophthora cactorum* works its way down to the roots, rotting them and killing the plant.

Phytophthora cactorum is also one of the causal agents of black rot of orchids. When infecting orchids, this organism first produces small black lesions on the pseudo bulbs of the orchid, which then enlarge and may engulf the entire pseudobulb, leaf, or move through the rhizomes to other portions of the plant prior to killing it. Diagnosis of orchid black rot by *P. cactorum* is through the identification of lemon shaped zoosporangia with either a papilla or a short pedicel, the presence of oospores, or molecular identification. Since there are multiple species of *Phytophthora* that are capable of causing disease on orchids, classification only to the genus level is required for proper prescription of disease management techniques.

Crown rot or root rot of strawberries is a common example of diseases of fruit crops caused by *P. cactorum*. On strawberries, *P. cactorum* infects the roots and the base of the plant causing stunting and reduced leaf size, with possible complete plant collapse later in the season. Similarly to apple trees, crown rots of strawberries caused by *P. cactorum* can be partially diagnosed by cutting the crown of the plant and observing brown vascular tissues, and root rots by brown or black stunted roots.

Life Cycle

Phytophthora cactorum is a homothallic (only having one mating type, can mate with itself) oomycete, and displays the right angled mycelial branching with a constriction at the base of the branch, which is highly characteristic of other *Phytophthora* genera. Within the hyphae, they have singular diploid nuclei that are regularly spaced. In addition, young hyphae only have cross walls separating reproductive parts; however older hyphae may have cross walls anywhere. Furthermore, although the hyphae are not the main survival unit of *P. cactorum*, as long as they are not completely desiccated, they are capable of surviving until just above freezing temperatures.

Phytophthora cactorum produces one sexually produced survival spore called an oospore, and one asexually produced survival spore called a chlamydospore. Oospores are double walled and uninucleate during dormancy, but become multinucleate in preparation for germination. In contrast, the chlamydospores have only one resistant wall and are multinucleate at all stages. Chlamydospores are larger than oospores in size, and are only formed under certain environmental triggers. The trigger for preferential formation of chamydospores over oospores can be either large temperature or moisture oscillations.

In addition to the chlamydospore, *P. cactorum* also produces another asexual spore called a sporangia. A sporeangia is a multinucleate dispersal structure with a thin wall and papilla that is formed on a sporeangiophore. Although the size may vary based on the environmental conditions in which they are formed, the width of a sporeangia is always more than 2/3 times than its length. Depending on moisture conditions, sporangia can either germinate or release zoospores.

Zoospores are produced in wet conditions by either oospores or sporangia. *Phytophthora cactorum* zoospores, are uninucleate, laterally biflagellate, and pear or lemon shaped. After being released, zoospores swim to a nearby wound on a suitable host, germinate, and enter wounds to cause a hyphal infection of the roots or crown vascular system.

Environment

Because *Phytophthora* is a soilborne pathogen, the ideal condition for *P. cactorum* growth is in saturated soil. *P. cactorum* stays in the soil as dormant resting oospores and chlamydospores, or within infected plant tissue. When conditions are met and the soil is wet enough, sporangia are produced, carrying on the life cycle of the pathogen. The minimum amount of time the plant must be saturated to produce an infection depends on factors such as genetics, physiological processes, and the environment. However, when a plant is allowed to sit in soil that is heavy and soggy for long periods of time, the chance of infection is increased. A plant's inability to fight off the pathogen is impeded when saturated soil conditions limit the amount of available oxygen for its roots. In many cases, most host plants are the most susceptible to infection during spring and autumn when the soil is wetter and at a more ideal temperature for zoospore production and activity.

Importance

P. cactorum was first described in 1870 as a cactus pathogen. Since then, it has been found to not only infect cacti, but a wide range of plants worldwide. *P. cactorum* was first reported in the United States in 1858 when infected apple trees were discovered in Michigan. By 1928, it had spread to Canada in the Okanagan Valley, British Columbia. Since then, it has caused around C$2 million in damage per year.

The importance of this oomycete is its vast host range and the damage it causes to major crops. This pathogen can cause root rot that stunts the host's growth and damages vascular tissue, which is especially detrimental to pear and apple orchards. It can also infect strawberry plants and cause crown rot, root rot, and leather rot of the fruit. This pathogen causes millions of dollars in damage, and disease management such as soil fumigation is also expensive.

Disease Management and Control

The best way to manage *P. cactorum* is by implementing an integrated management plan. The combination of soil fumigation and proper cultural controls will be the best option for plant health. Recommended chemical controls products include fosetyl Al, metalaxyl, and etridiazol. Prevention and sanitation are crucial because this pathogen is usually transmitted through cuts or injuries on the plant. The spores are easily transported through irrigation water and will splash to nearby plants. Elevating your plants above the ground can help to prevent infection. The pathogen thrives in moist soils therefore it is important to avoid very saturated soils as much as possible and one should work to prepare their soil for adequate drainage during the wet periods. Soil drainage and low soil pH may help to reduce the disease. Fertilizer regimen methods have been used to control against P. *cactorum*. These fertilizers include organic materials that release ammonia, nitrous acid and amendments to reduce the pH to less than 4. The use of raised beds and a carefully managed drip irrigation system will be important cultural practices that can be implemented. There are some form of Biological controls that have been found to be somewhat successful with *Enterobacter aerogenes* and *Trichoderma*.

Pathogenesis

When able to culture the *P. cactorum on* different media, it will produce a maple wilt associated phytotoxin. These phytotoxins cause browning of veins, desiccation of apple leaves, and wilting of tomato cuttings. Studies have shown that this the *P. cactorum* phytotoxin is hydrophilic in nature and will not move to the organic solvents. The chemical properties of this toxin seem to be similar to those of other *Phytopthora* species. Within the plant, the *P. cactorum* phytotoxin is most likely xylem transported through and can accumulate in the space in between leaves where it causes desiccation and withering.

In pathogenicity tests on strawberry plants in a greenhouse, the fruit isolates caused little disease, while strawberry root system isolates were highly aggressive. *P. cactorum* phytotoxin is thought to have a role in disease development and pathogenesis, however further study is required.

References

- "Orchid Research Program". Michigan State University. Archived from the original on 22 September 2012. Retrieved 17 October 2012
- HU J. S.; FERREIRA S.; WANG M.; XU M. Q. (1993). "Detection of Cymbidium mosaic virus, Odontoglossum ringspot virus, tomato spotted wilt virus, and potyviruses infecting orchids in Hawaii". Plant Disease. 77 (5): 464–468. doi:10.1094/PD-77-0464
- Rehner, Stephen A., and Francis A. Uecker. "Nuclear Ribosomal Internal Transcribed Spacer Phylogeny and Host Diversity in the Coelomycete Phomopsis." Canadian Journal of Botany, 1994, 72 (11), 1666-1674
- Pearson, M. N.; Cole, J. S. (1 November 1986). "The Effects of Cymbidium Mosaic Virus and Odontoglossum Ringspot Virus on the Growth of Cymbidium Orchids". Journal of Phytopathology. 117 (3): 193–197. doi:10.1111/j.1439-0434.1986.tb00934.x

- Wen, X., Q. Xu, Q. Cao, and X. Deng. "Promising Genetic Resources for Resistance to Powdery Mildew in Chestnut Rose (Rosa roxburghii) and Its Relatives in China." New Zealand Journal of Crop and Horticultural Science 34.2 (2006): 183-88. Print
- Verchot-Lubicz, J.; Ye, C.-M.; Bamunusinghe, D. (1 June 2007). "Molecular biology of potexviruses: recent advances". Journal of General Virology. 88 (6): 1643–1655. doi:10.1099/vir.0.82667-0
- Plich, Miroslawa; Rudnicki, Ryszard M. (March 1979). "Studies of the Toxins of Phytophthora cactorum Pathogenic to Apple Trees". Journal of Phytopathology. 94 (3): 270–278. doi:10.1111/j.1439-0434.1979.tb01559.x. ISSN 0031-9481
- Inouye, N (2008). Viruses of orchids : symptoms, diagnosis, spread and control. Netherlands: Blue Bird Publishers. p. 176. ISBN 9789079598038
- Wu, Xiang-Ming. "Effects of Temperature on the Length of the Incubation Period of Rose Powdery Mildew." European Journal of Plant Pathology 105.1 (1999): 13-21. Print
- HU J. S.; FERREIRA S.; XU M. Q.; LU M.; IHA M.; PFLUM E.; WANG M. (23 April 1994). "Transmission, movement, and inactivation of cymbidium mosaic and odontoglossum ringspot viruses". Plant Disease. 78 (6): 633–636. doi:10.1094/PD-78-0633
- Friskop, Andrew; Markell, Sam; Gulya, Tom (2009). "Downy Mildew of Sunflower" (PDF). Sunflower Publications. IPM Publications. Fargo, North Dakota: North Dakota State University Extension Service. Retrieved 4 March 2016
- Kubo, K. S.; Freitas-Astúa, J.; Machado, M. A.; Kitajima, E. W. (8 May 2009). "Orchid fleck symptoms may be caused naturally by two different viruses transmitted by Brevipalpus". Journal of General Plant Pathology. 75 (3): 250–255. doi:10.1007/s10327-009-0167-z
- Gubler, W. D.; Davis, U. C.; Koike, S. T. "Powdery Mildew on Ornamentals". University of California- Agriculture and Natural Resources. University of California Statewide Integrated Pest Management Program. Retrieved 3 December 2015

Permissions

All chapters in this book are published with permission under the Creative Commons Attribution Share Alike License or equivalent. Every chapter published in this book has been scrutinized by our experts. Their significance has been extensively debated. The topics covered herein carry significant information for a comprehensive understanding. They may even be implemented as practical applications or may be referred to as a beginning point for further studies.

We would like to thank the editorial team for lending their expertise to make the book truly unique. They have played a crucial role in the development of this book. Without their invaluable contributions this book wouldn't have been possible. They have made vital efforts to compile up to date information on the varied aspects of this subject to make this book a valuable addition to the collection of many professionals and students.

This book was conceptualized with the vision of imparting up-to-date and integrated information in this field. To ensure the same, a matchless editorial board was set up. Every individual on the board went through rigorous rounds of assessment to prove their worth. After which they invested a large part of their time researching and compiling the most relevant data for our readers.

The editorial board has been involved in producing this book since its inception. They have spent rigorous hours researching and exploring the diverse topics which have resulted in the successful publishing of this book. They have passed on their knowledge of decades through this book. To expedite this challenging task, the publisher supported the team at every step. A small team of assistant editors was also appointed to further simplify the editing procedure and attain best results for the readers.

Apart from the editorial board, the designing team has also invested a significant amount of their time in understanding the subject and creating the most relevant covers. They scrutinized every image to scout for the most suitable representation of the subject and create an appropriate cover for the book.

The publishing team has been an ardent support to the editorial, designing and production team. Their endless efforts to recruit the best for this project, has resulted in the accomplishment of this book. They are a veteran in the field of academics and their pool of knowledge is as vast as their experience in printing. Their expertise and guidance has proved useful at every step. Their uncompromising quality standards have made this book an exceptional effort. Their encouragement from time to time has been an inspiration for everyone.

The publisher and the editorial board hope that this book will prove to be a valuable piece of knowledge for students, practitioners and scholars across the globe.

Index